Learn Math Fast System

Volume 6

Applications of Algebra

By JK Mergens

ISBN: 9780984381470
Presidential $1 Coin images from the United States Mint

CONTENTS

INTRODUCTION

Welcome to Volume VI of the *Learn Math Fast System*. For best results, read Volumes I - V first, as this book assumes you know how to solve algebraic equations.

At times, I use color ink in this book. For example, I might say look at $-8x^2$ in the equation below.

$$x^2 + 8x - 8x^2 = 0$$

The red color is there to help you locate the term I am talking about. Other times I may use two colors to help you see the Like Terms, like this:

$$3a + 4x^2 + 5a - 4y^2 + 9x^2$$

Use the colors to help you follow along with what I'm saying.

I encourage you to read each lesson several times. Spend a week or more learning each lesson thoroughly. Read each example until you can solve the problem completely on your own with ease. If you get stuck on a problem, use the answer key to help you out.

After you finish the book, go back and try to solve each story problem again, until you can complete each one without peeking at the answers. Then go online and take a practice SAT test. There are several different ones available, take them all. It will get easier each time.

The title says *Learn Math FAST*, so let's get started.

CHAPTER 1: BUILDING EQUATIONS

LESSON 1: FINDING THE KNOWN AND UNKNOWN

Now that you can solve nearly any algebraic equation and you know a handful of useful formulas, let's use them to help us solve story problems. As it turns out, algebra isn't completely useless. It can be used to help figure out problems that would otherwise be very challenging to solve.

Most algebra story problems look and sound very difficult or even impossible to solve. But once you learn how to spot the clues to help you build an equation, it becomes as simple as algebra!

Let me explain what I mean. Read the story problem below. I'll explain how to solve it, next.

In order to get an average score of 91, Josh needs to receive a total of 273 points on three tests. He received scores of 92 and 87 on his first two tests. What score must he get on the third test to make sure he will get an average score of 91?

In order to solve the problem above, we need to create an algebraic equation. Then we can use our algebra skills to solve the problem.

The first clue to look for in every story problem is the *unknown*. It's easy to find the unknown because it will usually be in the form of a question. Look for the *question* in the story problem above. It starts with the words "*What score.*" That is the unknown number we are looking for; the score Josh needs to get on test #3. Since we don't know that score yet, we need to create a variable that will stand for his score on test #3. Let's call that unknown score "x."

$$x = \text{score from test \#3}$$

Now that we have found the unknown and labeled it "x," we look for the *known* numbers. Here is what we *know* from the story problem on the last page.

92 = score on test #1
87 = score on test #2
273 = total score for all 3 tests

Do you remember how to find the average of 3 numbers? You add up the 3 scores and then divide by 3, to get the average score. We know that we want all 3 test scores to equal 273 because when you divide that number by 3, you get an average score of 91. So let's do that. Let's make an equation that shows all 3 scores totaling 273.

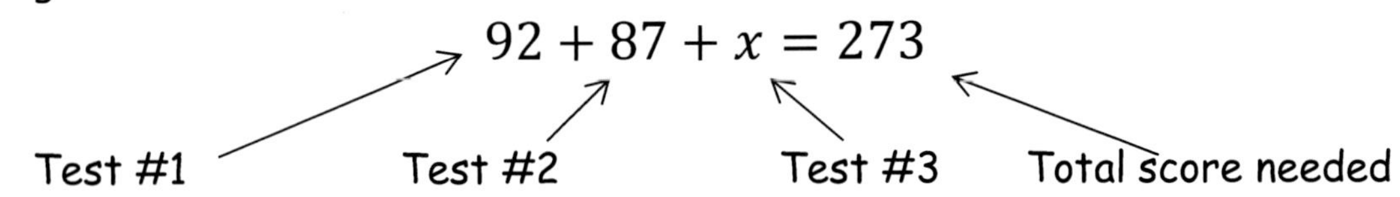

Now we just do some basic algebra. Add these two numbers together.

$$92 + 87 + x = 273$$

$$179 + x = 273$$

Subtract this number from both sides, to get x by itself.

$$x = 273 - 179$$
$$x = 94$$

Josh needs to get a score of 94 on his third test, in order to get an average score of 91.

OK, that was an easy one, so let's try that again with a more challenging problem. Read the next story problem. The clues are in color. The known values are in red and the unknown value is written in blue. Next, I will help you solve it.

Amanda is saving money to go to college in the fall. She already has $300, but she needs a total of $2,400 to pay for her classes. She earns $14 per hour at her job. How many hours does Amanda need to work, in order to save a total of $2,400?

We need to build an equation to solve this. First step: find the unknown number and call it "x." The unknown value is "*How many hours Amanda needs to work.*" That is our "x."

$$x = \text{hours worked}$$

Next, I will list all the values that we do know:

300 = money she already has
2400 = total amount needed
14 = amount she will earn for each hour (x).

Let's use all that information to build an equation. Remember, an equation is a math sentence that says, "this EQUALS that." We want Amanda's money to equal $2,400, so start there. We want $2400 to equal something.

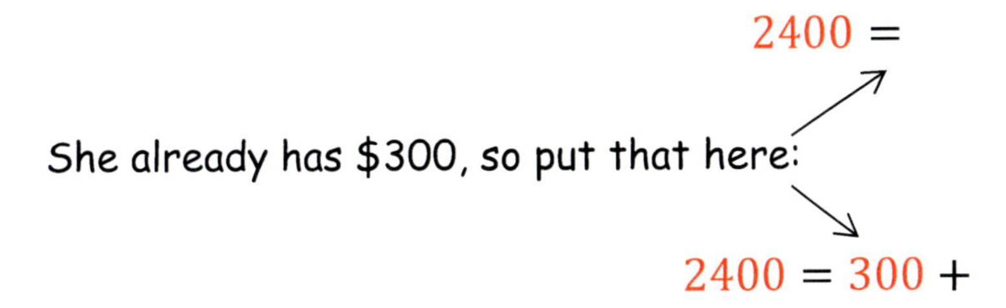

She will earn $14 for every hour she works. If she works for 2 hours, she will earn $14 times 2 hours. If she works for 10 hours, she will earn $14 times 10. In our math equation, we will call that "14x" because she will work for "x" number of hours, times $14. Add 14x to the equation below.

$$2400 = 300 + 14x$$

Do you understand the equation that we have built above? It is saying that

Amanda needs $2400 total. She already has $300 and she can earn $14/hour at work.

$$2400 = 300 + 14x$$

We are trying to figure out HOW MANY hours she needs to work to make this equation true. Let's solve for x to find out.

$$2400 = 300 + 14x$$

Subtract 300 from both sides

$$2400 - 300 = 14x$$

$$2100 = 14x$$

Divide both sides by 14, to get x by itself

$$\frac{2100}{14} = \frac{14x}{14}$$

$$150 = x$$

Our equation tells us that Amanda needs to work 150 hours in order save up enough money for college. Does that make sense? You can find out if that is the right answer by replacing the "x" in our equation with "150." If $14 times 150 hours, plus $300, equals $2400, then we have the right answer. Here is another example.

Patty received a rebate check. She earns points every time she spends money at the dentist's office. She spent a total of $4000 and they sent her a check for $25. What percentage did the dentist office use to calculate her rebate?

What are we trying to figure out? A percentage; 25 is what percent of 4000? To find a percentage of a number, we multiply. 4000 times some number will equal 25, right? Let's write that out.

$$25 = 4000x$$

The first step is to get x by itself. 4000 is MULTIPLIED by x, so we need to do the opposite, DIVIDE.

$$\frac{25}{4000} = \frac{4000x}{4000}$$

$$\frac{25}{4000} = x$$

$$25 \div 4000 = .00625$$

$$x = .00625$$

OK, what does that mean? We are looking for a percentage. Our answer is a number, not a percentage. Use your percentage skills to change this number into a percentage.

The answer is .625%. The dentist will give back .625% of every dollar spent at his office. That's less than 1 percent.

Let's try one more together. Read the story problem below.

A baseball team held a fundraiser inside the mall to buy new uniforms. The team raised $1250, but they have to pay $350 to rent the mall space. Each new uniform costs $60.00. How many uniforms will the team be able to buy?

What is the first step? That's right! Look at the question to find out the unknown number. The question in the story problem above is asking "how many uniforms," so that is the unknown number. We will call that number "x."

x = number of uniforms

Now let's build an equation around x. Each uniform cost $60, so the total amount needed is 60x. Get it? $60 times the number of uniforms is how much money the team needs. That is the first part of our equation.

$$60x =$$

We want the amount they need to equal the amount they have. So what do they have? The team raised $1250, but they need to pay the rent, so be sure to subtract that amount.

$$60x = 1250 - 350$$

Now we can do the math. First, let's do the subtraction problem.

$$60x = 900$$

Next, divide each side by 60, to get x by itself.

$$\frac{60x}{60} = \frac{900}{60}$$

$$x = 15$$

With the money the team raised, they can buy 15 uniforms at $60 each.

Some story problems will require you to dig through your math tool box to pull out one of those formulas you learned in geometry class. For example, if a story problem is referring to a right triangle and you are to find the length of the hypotenuse, you would want to fill in the Pythagorean Theorem with the information given. Let's use this next story problem as our example.

Adam is trying to build a bike ramp. The ramp will be 6 feet tall and the base of the ramp will be 8 feet long.

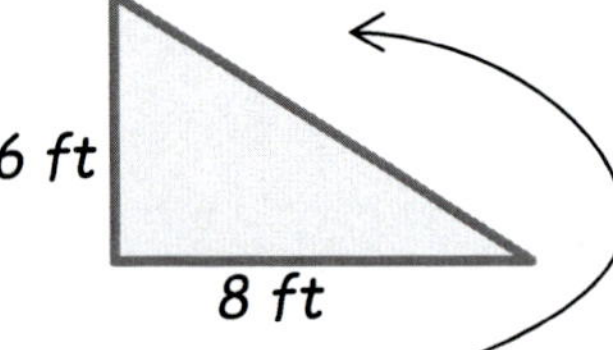

What will be the length of ramp's tread?

This story problem is asking us, "How long is the ramp?" But to a math person, the question is, "What is the length of the hypotenuse in the right triangle formed?"

To find the length of the hypotenuse, we use the Pythagorean Theorem.

$$a^2 + b^2 = c^2$$

Fill in the formula above with the information we already know. Side A is 6 feet long and Side B is 8 feet long.

$$6^2 + 8^2 = c^2$$

Solve the equation above to find the length of the ramp.

$$\begin{gathered} 6^2 + 8^2 = c^2 \\ 36 + 64 = c^2 \\ 100 = c^2 \\ \sqrt{100} = \sqrt{c^2} \\ 10 = c \end{gathered}$$

The length of the ramp is 10 feet.

Try building and solving equations on the next worksheet. Remember to first find the unknown number and call it x. Then find the known numbers to help you build an equation that makes sense. If a problem asks you to find the area, volume, or circumference, be sure to use those formulas to help you create an equation.

WORKSHEET 1

Name____________________________________ Date ____________

1. The auditorium can only hold 450 people. There are already 45 people there, working to put on a play for their family and friends. Each person working on the play wants to sell tickets. How many tickets can each person sell without exceeding the total number of people allowed in the building at one time?

   What is the unknown? ____________________ call it "x"

   Total number of people allowed? ______________

   Number of people in the play? _____________

   Total People Allowed   People already there   Total tickets

   _________________ - ______________ = 45x

2. You are given 128 feet of chain link fence. It is your job to fence off a square shaped pen for several dogs. The pen must be 1024 square feet, in order for each dog to be comfortable. What should be the length of each side of the pen?

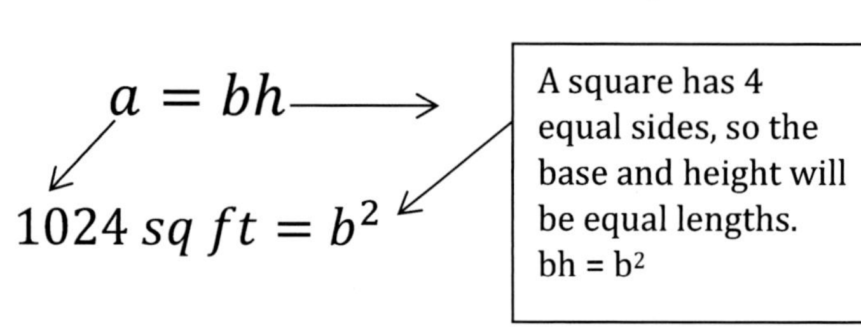

Worksheet 1 page 2

3. An artist painted a big picture on the side of a building. The picture is in the shape of a circle with an area of 113.04 square meters. What is the diameter of the painted circle?

$$area\ of\ a\ circle = \pi r^2$$

$$\underline{\hspace{3cm}} = \underline{\hspace{3cm}} \times r^2$$

4. There is a lot of snow on the roads. A truck driver needs to put chains around his tires, so his truck doesn't slip in the snow. The tires measure 30 inches, from one side to the other. How long do the chains need to be, in order to go around the tires?

5. There are 20 cubic feet of sand left over from a project. Mr. Wilson wants to build a sand box to hold the left-over sand. He built a rectangular box. The box is 5 feet long by 2 feet wide. How deep will the sand be once it is poured into the sandbox?

LESSON 2: DISTANCE DIVIDED BY RATE EQUALS TIME

The last worksheet had story problems that involved formulas such as Area, Circumference, and Volume. Next you will learn a new formula and then I will show you how to use it in a story problem. It is written below.

$$\frac{distance}{rate} = time$$

Let's take a few minutes to learn about the very useful formula written above. It is the "Distance Over Rate Equals Time" formula, but let's just call it the "Distance Formula" for short. This formula can be used to figure out three different things. You can find out how fast something traveled (rate), how far it traveled (distance) or how long it took to get there (time). As long as you know two of the numbers; distance, time, or rate, you can figure out the third number with this formula.

But before we try to use this formula, let's take a closer look to get a little more familiar with it first. I will use the letters, "d, r and t" to represent, "Distance, rate, and time." Now look at this equation.

$$\frac{d}{r} = t \qquad \frac{15}{5} = 3$$

That equation says, "Fifteen divided by 5 equals 3." Our new formula is just like that equation. Our new formula is saying, "Distance divided by rate equals time."

$$\frac{d}{r} = t$$

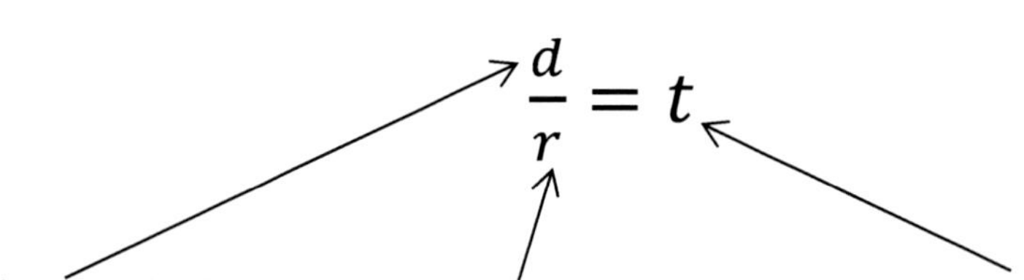

The letter "d" stands for Distance, "r" stands for rate, and "t" is for time." *Rate* is a fancy word for speed, you can use either word. Sometimes it is even called "Rate of Speed." Now look at the equation below. I switched around the 3 and the 5 this time.

$$\frac{15}{3} = 5$$

Now this equation says, "Fifteen divided by **3** equals **5**." This should be no big surprise to you. You already know this is true. Now I'll do the same switch-a-roo with our new formula.

$$\frac{d}{t} = r$$

I switched the time and rate, but it is still true. Just like when I switched the 3 and the 5 around. Now it reads, "Distance divided by time equal rate." Now look at this equation.

$$15 = 3 \times 5$$

Of course, you know this is true. Now I'll do the same with our formula.

$$Distance = rate \times time$$

$$d = rt$$

This is also true, however, that makes it very difficult to remember which way it goes. So, I have a way to make it easier for you to recall. The "Distance" Formula will always start with the word "Distance." The "Distance" is always divided by either the "rate" or "time." For example, you can say "Distance divided by Rate equals Time" or "Distance divided by Time equals Rate." Either one is correct; just be sure to start with the word "Distance" and then divide by either rate or time.

Basically, if you are trying to solve for the time, then you would want to go with "Distance divided by rate equals time." And if you were trying to solve for the rate, then it would be quicker use, "Distance divided by time equals rate." That will save you a step when you are trying to solve your equation.

Now let me show you how handy this formula can be. I will start off with a simple example.

Let's say you drove a car at the same speed for 1 hour. After one hour, you had traveled 60 miles. What was the speed of the car? In this situation, "Distance" is how far you traveled. The "Rate" is how fast you drove, and "Time" is how long you were driving. So, let's see, the distance is 60 miles, the time is one hour and the rate is unknown. I'll fill in the Distance Formula with our information. We don't know the rate, so that will be "x."

I chose to go with "Distance divided by Time" to save myself a step. (If I had used Distance divided by Rate, then I would just end up having to do the el switch-a-roo with rate and time to end up back at Distance divided by Time.)

$$\frac{Distance}{time} = rate$$

$$\frac{\text{60 miles}}{1\ hour} = \text{x}$$

Now that we have built an equation, we can use algebra to solve it. But wait, it's already solved; x is already by itself. The answer is 60 miles per hour. The rate the car drove is 60 miles per hour. That was easy, so let's try another one.

How long will it take a train to travel 300 miles, if it is going 80 mph?

Fill in the formula with the values (numbers) that are known and put a "t" in the place of the unknown number. I'll use Distance divided by Rate, this time.

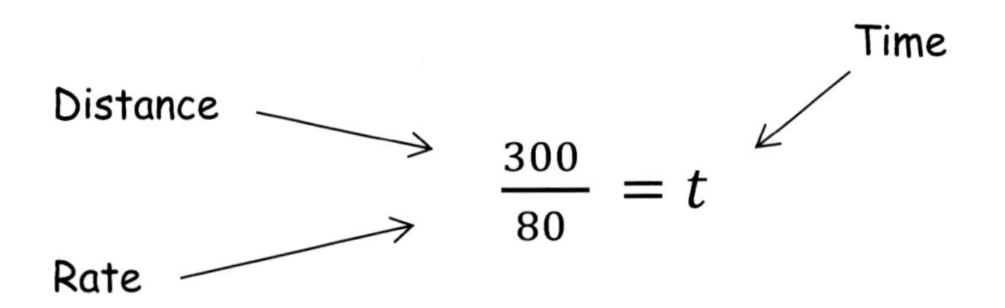

$$\frac{300}{80} = t$$

$$\frac{300}{80} = t$$

30 ÷ 8 = 3.75

$$\frac{30}{8} = t$$

$$t = 3.75\ hours$$

Here is another one.

How fast did Courtney drive home in the snow? She traveled 10 miles, but it took her 2 ½ hours.

Well, let's see. We know she traveled 10 miles and we know how long it took her. The unknown value is her rate of speed. I'll fill in the Distance Formula with the values we know. This time I used Distance divided by Rate…and watch what happens.

$$\frac{10\ miles}{x} = 2.5\ hours$$

I have to do the el switch-a-roo, to switch around the rate and time, then I can solve it. I should have gone with "Distance divided by Time" in the beginning.

$$\frac{10}{x} = 2.5$$

I have switched them around below. Now you know the value of x.

$$\frac{10}{2.5} = x$$

But what does that mean? How does that tell us Courtney's speed? Let's put the units back into this equation.

$$\frac{10\ miles}{2.5\ hours} = x$$

That reads, "10 miles per 2.5 hours." Remember, this line means "per." Now let's reduce that fraction, but keep the "miles per hour" unit.

$$10\ miles \div 2.5\ hours = 4\ miles\ per\ hour$$

Courtney drove 4 miles per hour. Sheesh! That was a slow drive.

Now let's solve for the "Distance" part of our formula.

A cougar was seen running at 40 miles per hour. He continued to run at that speed for 15 minutes. How far did he get?

This time we have the speed and the time, so we need to solve for the distance.

$$\frac{x}{40\ miles/hour} = 15\ minutes$$

Mmm...as I look at this equation, I notice that we have both hours and minutes. Our units of time should be the same. I can write "15 minutes" as ".25 hours," instead (15 minutes is $\frac{1}{4}$ an hour; $\frac{1}{4}$ =.25). Now our problem looks like this:

$$\frac{x}{40} = .25$$

We could try the "el switch-a-roo," but that won't help us solve for x. The proper thing to do is multiply both sides by 40, that will get x by itself.

$$\frac{x}{40} = .25$$

$$x = .25 \times 40$$

Oh look! That's the same thing as "Distance equals time x rate." I should have started there; I could have saved a step.

$$x = .25 \times 40$$

$$x = 10\ miles$$

The cougar ran for 10 miles. I don't know if that's true, though. Can a cougar really run that fast for 10 miles?

Try solving some story problems using the Distance Formula on the next worksheet.

Even though I could have saved myself a step by starting with "Distance equals Rate times Time," I still like to start with "D/R = T" because it's easier to remember. My point is to try and set up your equation so the "equals this" part is also your unknown number. But if you don't, that's OK, you can just stick with "Distance divided by rate equals time" and if you have to do an extra step…oh well, no big deal.

WORKSHEET 2

Name ______________________________________ Date ______________

1. It took 3 hours for the airplane to travel 1800 miles. What was the airplane's rate of speed?

2. Teresa ran a 20-mile marathon. Her average speed was 5 miles per hour. Approximately how long did it take her to finish the race?

3. John and Pat drove from Seattle to Portland. They drove an average of 50 miles per hour and it took about 3 hours to get there. How far did they travel?

4. A golf ball travels 440 yards. It flew through the air at 100 miles per hour. How long did it take to land?

LESSON 3: DISTANCE FORMULA IN STORY PROBLEMS

Now that you understand how to use the Distance Formula, let's bump it up a bit, so it's a little more Algebra "Two-ish."

Most upper level math tests, like the SAT or college placement tests, will give you a story problem that involves two trains going opposite directions. They sound nearly impossible to solve, but once you realize how easy it is to build an equation, you will find these types of problems simple. Read the example below.

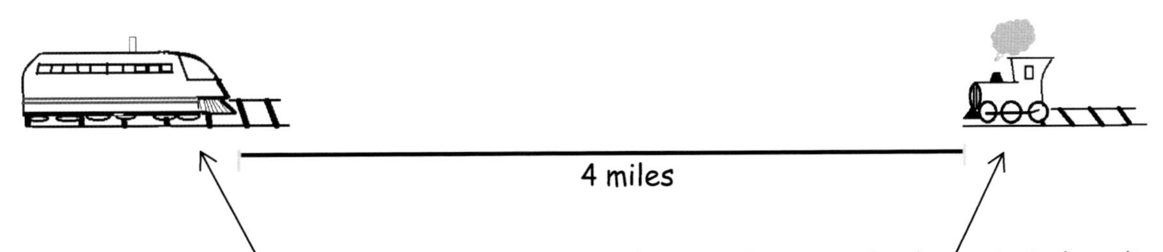

A commuter train is heading east at 50 miles per hour. A little train is heading west at 30 miles per hour. The trains are 4 miles apart. How long until they meet?

OK, you know this is a "Distance Formula" question, but there are two different rates, no time, and their distances are combined! Yikes!

We have to look at this problem logically. First, we will focus our attention on the little train. It is going to travel some UNKNOWN distance at 30 miles per hour. And we would like to know how long it will take this little train to get there.

OK, let's say the trains meet each other somewhere, about, oh I don't know...here.

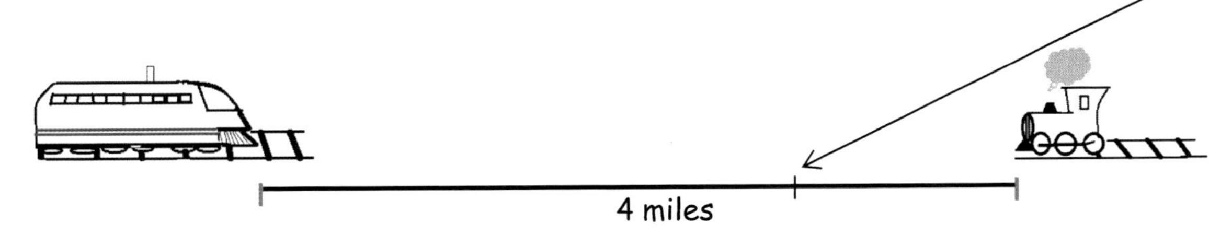

Then this distance is one of our unknown amounts. We will call it x.

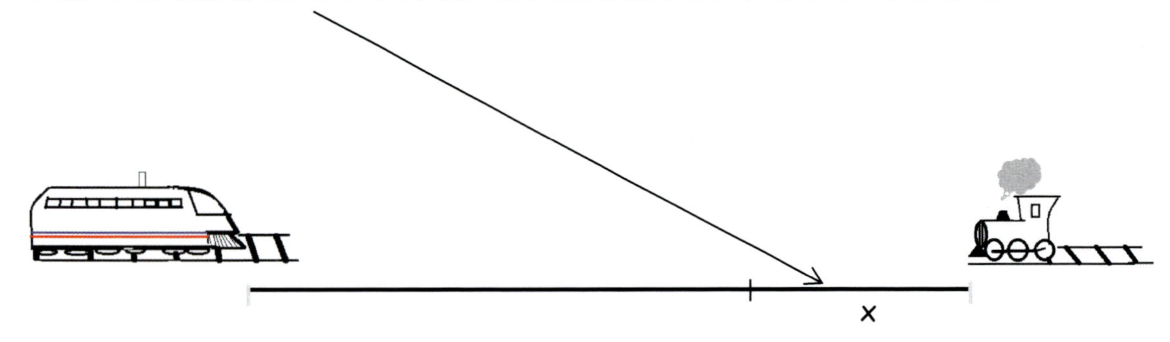

So, if the little train traveled "x" miles and the entire distance is 4 miles, then the commuter train will have to travel "4 miles - x miles." Do you understand that? Let's pretend the little train went 1 mile. Then the commuter train would go 4 - 1 miles. But we don't know how far the little train went, so we call it "x."

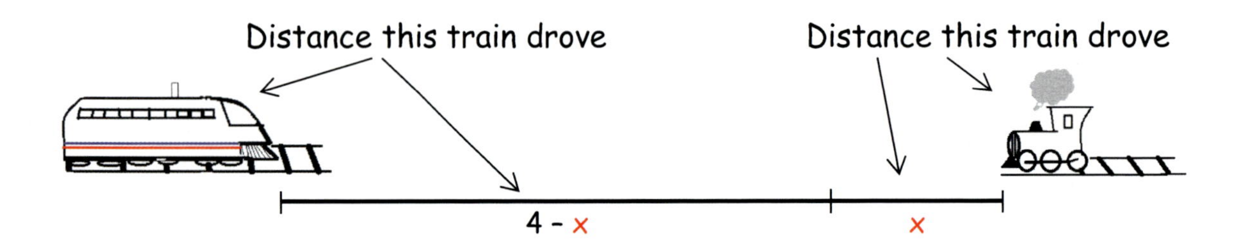

OK, now we have two distances and two rates. Let's make an equation for each train using the Distance Formula.

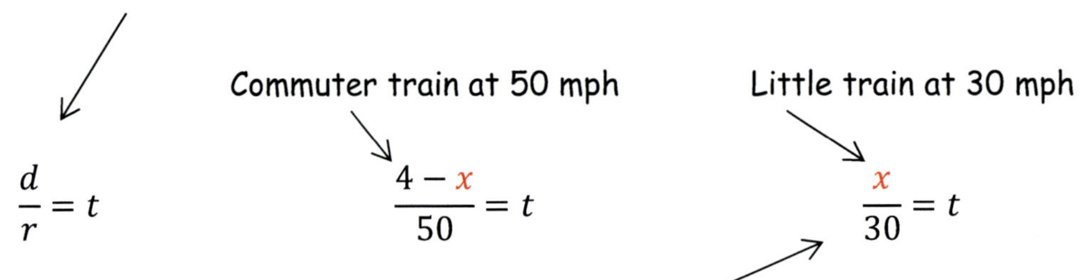

$$\frac{d}{r} = t \qquad \frac{4 - x}{50} = t \qquad \frac{x}{30} = t$$

Now comes the fancy part! Look at this equation here. Solve for x by multiplying both sides by 30. You get this, right?

$$x = 30t$$

Now we know how far the little train traveled. It traveled $30t$ miles. So that means that the big commuter train traveled $4 - 30t\ miles$.

Replace the "x" in the commuter train's equation with $30t$.

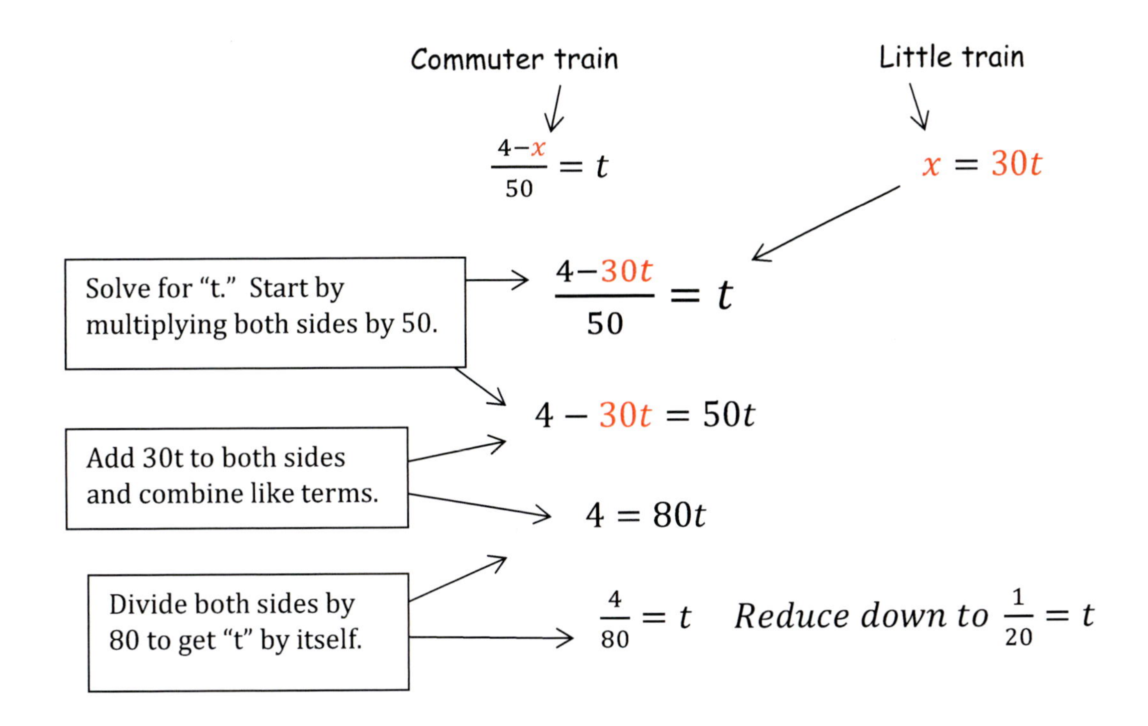

OK, so $t = \frac{1}{20}$, what does that mean? That is the time it will take for the two trains to meet each other, $\frac{1}{20}$ of an hour. But what is that? Well let's see, an hour is 60 minutes long, so we want to know how much is $\frac{1}{20}$ of 60 minutes. I'll change those words into math.

$$x = \frac{1}{20} \times \frac{60}{1}$$

$$x = \frac{60}{20} \text{ or } \frac{6}{2} \text{ or } 3 \text{ minutes}$$

Do you see how changing a story problem into an algebraic equation can help you solve for the answer? Let's try another one together. Read the following story problem.

Anita drives for 5 hours at some unknown speed and then she speeds up by 10 mph and drives for 3 more hours. She traveled a total of 430 miles. What two rates of speed did she travel?

The first step is to figure out WHAT we are looking for; the unknown. This question asks, "What two rates of speed…" so that is our "x." Anita drove x mph.

$$x = first\ speed$$

And how long did she drive that fast? She drove x mph for 5 hours. Then she sped up and drove 10 miles per hours faster for 3 more hours. That would be written as "x + 10 mph," wouldn't it? I'll draw a little picture, so this is easier to visualize.

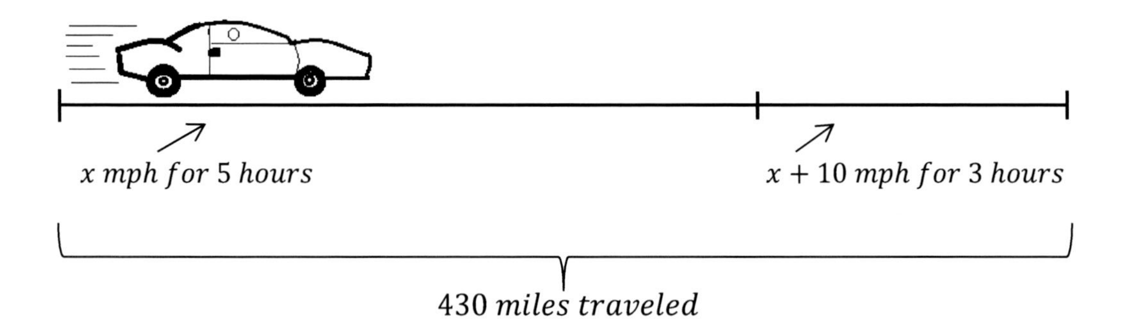

Let's pull out our distance formula and fill in some of the variables with our info.

$$\frac{d}{r} = t$$

Mmm...since we know the distance is 430 miles, let's rearrange this formula, so it says, "Distance equals something."

$$d = rt$$

That's more like it, now I can fill in the distance.

$$430\ miles = rt$$

OK, we've got our distance, now we need to fill in the "rate" and "time." The first portion of Anita's trip was driven at x mph for 5 hours. That is the rate and time.

$$430 = 5x$$

But don't forget about the second portion of her trip. She also drove for 3 hours at x + 10 mph. But think about that for a moment...how will you write that? If you wrote $3x + 10$, that would be, "3 hours at the original speed, plus 10 miles." That's not right. We want to write 3 hours TIMES $x + 10$.

$$430 = 5x + 3(x + 10)$$

Rub your palms together briskly, and let's do some algebra!

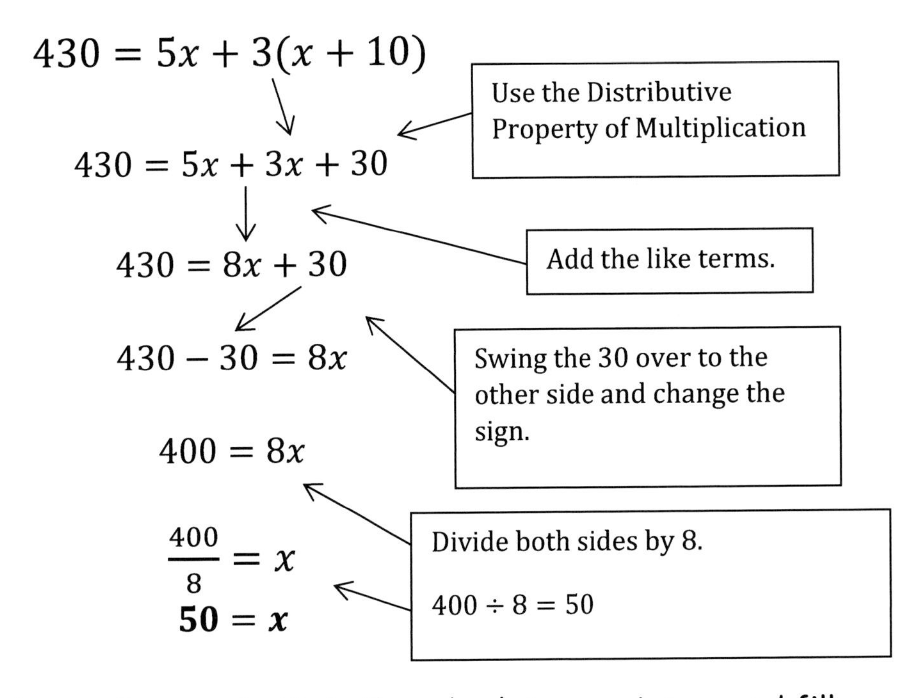

Tada! We have an answer for x. So now what? Let's go back to our picture and fill in the "x" with 50.

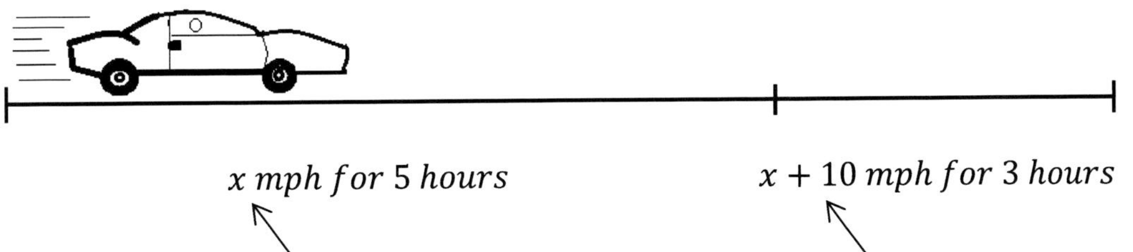

Anita traveled at 50 miles per hours for 5 hours and 60 mph for 3 hours.

There is one other way to use the distance formula. It isn't just for solving "traveling" types of problems; it can be used to solve other kinds of problems too.

For example, the "rate" doesn't have to be "miles per hour," it could be "number of lawns mowed per hour" or "number of books read per day." Here is an example of a "rate" story problem. Get it? Instead of a "distance" problem, I said "rate."

Esther can peel 4 potatoes per minute. How long will it take her to peel 68 potatoes?

Let's pull out the distance formula...ahem...I mean the RATE formula. (In case you don't get my humor, they are the same thing.)

$$\frac{d}{r} = t$$

In this situation the "Rate" is how fast Esther can peel potatoes. She can peel 4 per minute, $\frac{4}{1\ minute}$. That amount goes here.

Remember: This line means "per."

$$\frac{d}{\frac{4}{1}} = t$$

Don't be nervous about having a fraction in the denominator, it will be really easy to solve. We are looking for the "time" it will take to peel "68" potatoes. So you could say that peeling 68 potatoes would be "going the distance." And "time" is our unknown variable.

$$\frac{68}{\frac{4}{1}} = t$$

Do you know how to solve this equation? Since $\frac{4}{1}$ is the same thing as 4, just divide, $68 \div 4$.

$$\frac{68}{4} = t$$

$$17 = t$$

It will take Esther 17 minutes to peel 68 potatoes at the rate of 4 potatoes per minute.

Now let's say Esther gets some help. Her mom, Keisha, has a fancy automatic potato peeler that can peel 13 potatoes per minute. Now how fast can the two of them peel 68 potatoes together?

That's no big deal. I'll just add Keisha's rate and Esther's rate together.

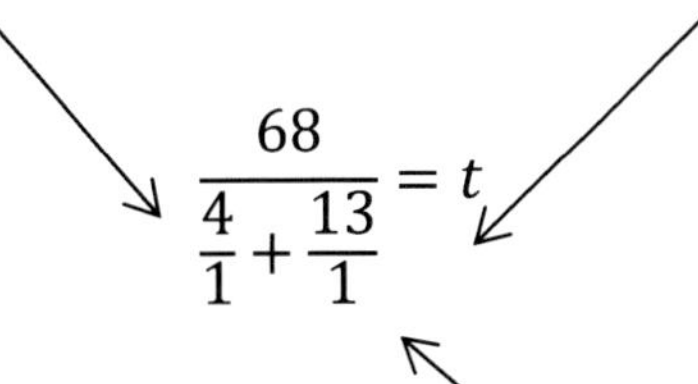

$$\frac{68}{\frac{4}{1}+\frac{13}{1}} = t$$

Before I do anything else, I'm going to add these two whole numbers together.

$$\frac{68}{17} = t$$

$$4 = t$$

It will take only 4 minutes for the two of them to peel 68 potatoes.

Complete the next worksheet. If you get stuck, try really hard to figure it out yourself. But if you just can't figure it out, peek at the answer to give yourself a little hint.

WORKSHEET 3

Name ______________________________________ Date ____________

1. Trinity walked to town at a rate of 4 mph. She rode the bus back at 40 mph. The town is 2 miles away. How long did it take Trinity to travel to town and back?

2. A dog is resting against a pole to which he is attached with a 24-meter chain. He watches a cat approach to within 8 meters of the pole and promptly gives chase. The dog runs at 12 meters per second and the cat runs at 10 meters per second. Which does the dog reach first, the cat or the end of the chain?

Worksheet 3 page 2

3. A pilot flew north against the wind, which was blowing at 40 mph, for 5 hours. Returning south at the same speed, with the wind pushing him 40 mph, he made the trip in 3 hours. What was the speed of the plane? (Hint: the distance is the same in both directions).

4. Two workers plan to mow 14 lawns. The older worker can mow 2 $\frac{1}{2}$ lawns per hour. The younger worker can mow 1 lawn per hour. How long is this going to take the two workers to mow all 14 lawns.

LESSON 4: CONGRUENT TRIANGLES

In this lesson, you will see one of the ways that geometry and algebra join forces. We are going to borrow some triangles from Geometry to help us solve story problems. First, we will look at some *Congruent Triangles.* When two triangles (or lines or angles) are the same size and shape, they are "congruent." The sides have to be the same length and the angles must be the same too, in order to be called congruent. Look at the two triangles below.

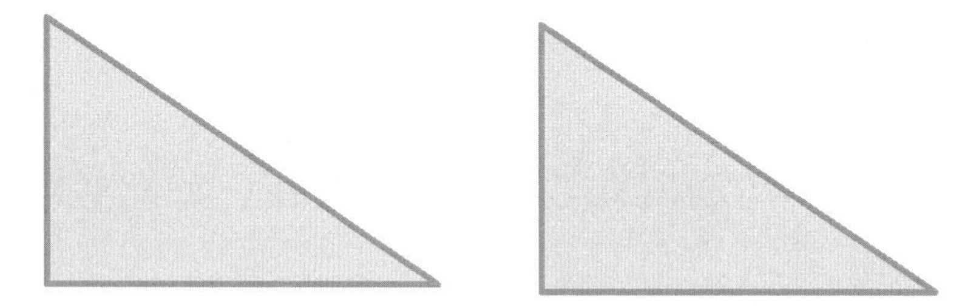

These two triangles are congruent. The sides are the same length and the angles are the same too. I know they are congruent because I copy/pasted it. Even if I turn one of the triangles around, they are still congruent.

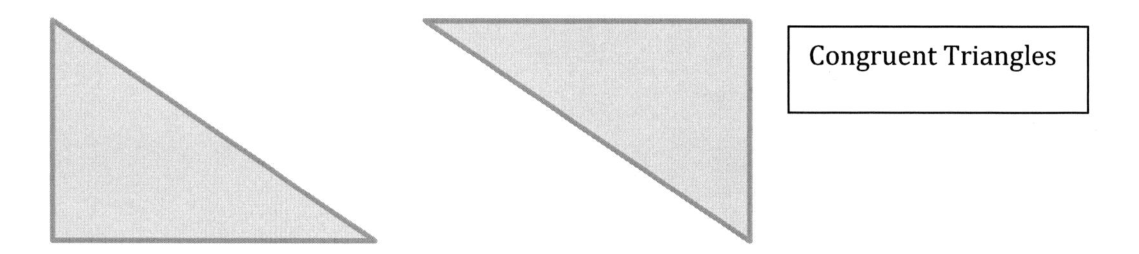

And even if they overlap, they are still congruent.

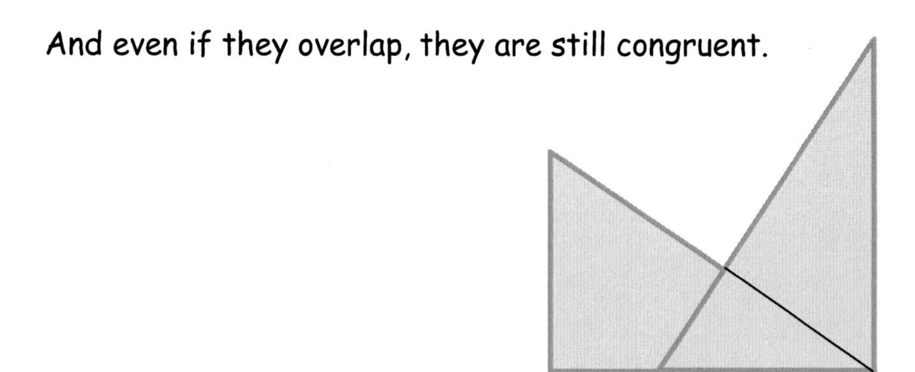

Let's straighten them back out and take a closer look at our two congruent triangles. We'll call them Δabc and Δdef.

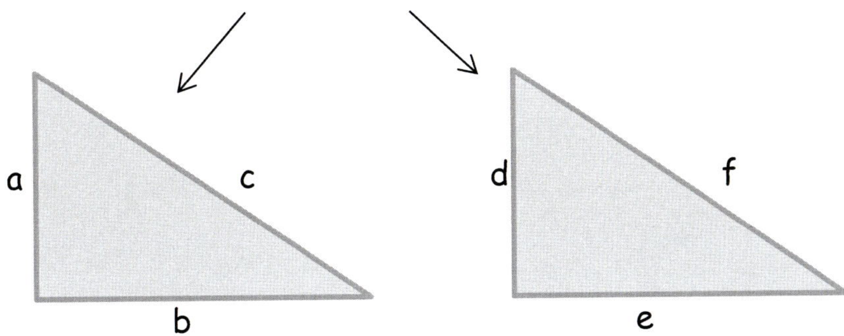

The math symbol for congruent is:

Since these two triangles are congruent, we write $\Delta abc \cong \Delta def$. You could also say that side a is congruent to side d, or $a \cong d$. It is also true to say $b \cong e$.

That's simple enough, right? Identical triangles have identical sides and angles. That makes sense to me. The two triangles below are NOT congruent. They are similar, but they are not the same size, so they are $\ncong$ (not congruent).

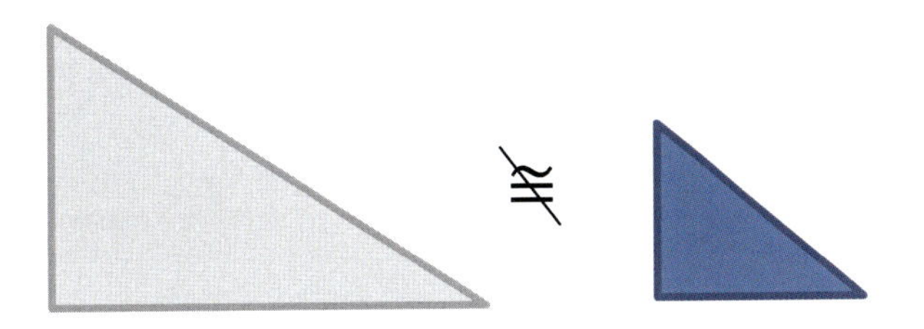

We will talk more about similar triangles later. For now, we will just discuss congruent triangles and then use them to solve some algebra problems.

We will start off very simple. This time, instead of giving each side of the triangles a letter, I will give each side a value. Look at these next two congruent triangles.

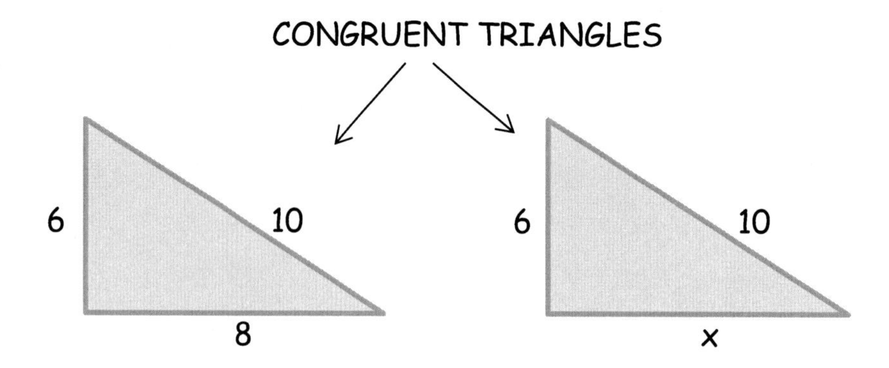

Just look at these two congruent triangles and then try to solve for "x." Since all sides are equal, I would guess that x = 8, wouldn't you?

OK, that was easy, so try this one. Solve for a, b, and c in the two congruent triangles below.

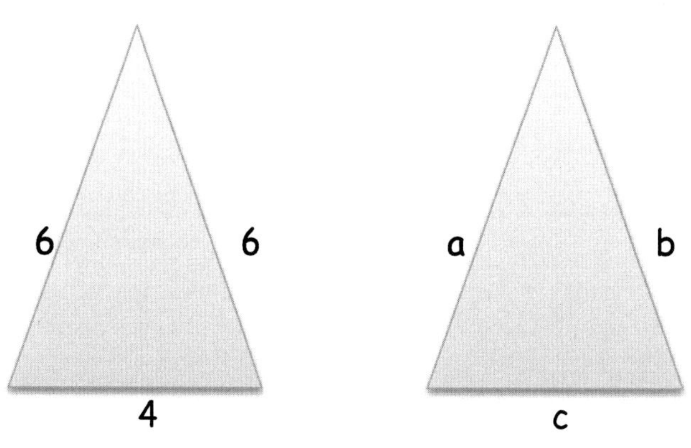

This one is easy too. Since all sides are congruent, a = 6, b = 6 and c = 4.

Sometimes triangles aren't lined up as nicely as the ones above. Sometimes they are twisted around making it more difficult to figure out which two sides are the congruent sides. So, mathematicians use little marks to quickly identify which two sides, lines, or angles are congruent. I put those little marks on the next two congruent triangles.

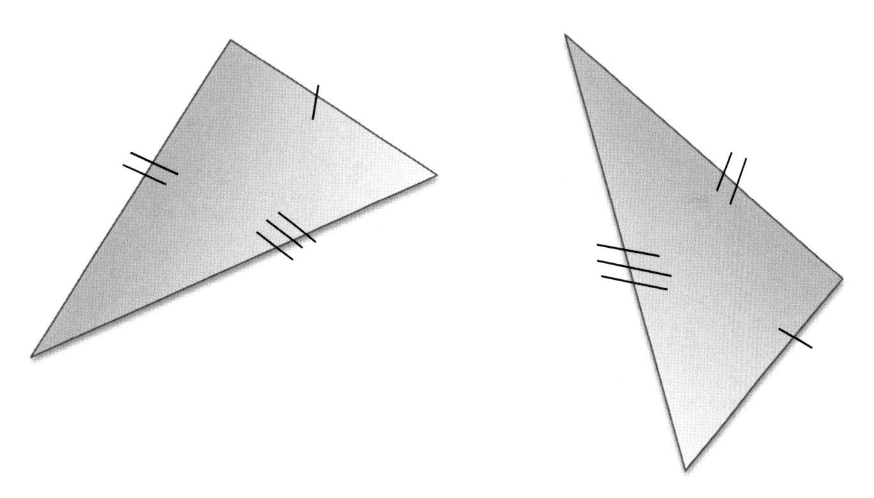

Do you see how each congruent side has matching marks? These same kinds of marks are used to show congruent (or equal) angles too. I've drawn those on the isosceles triangle below.

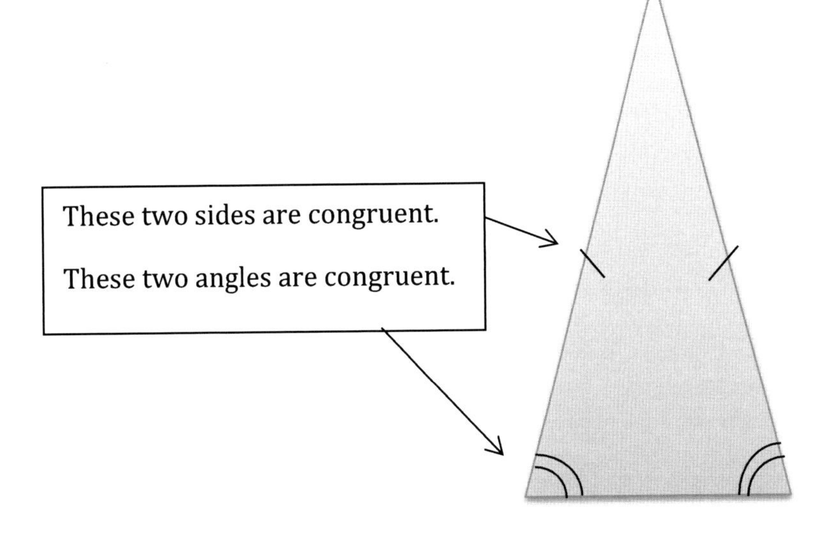

With that information in mind, try to solve for x in the next set of congruent triangles.

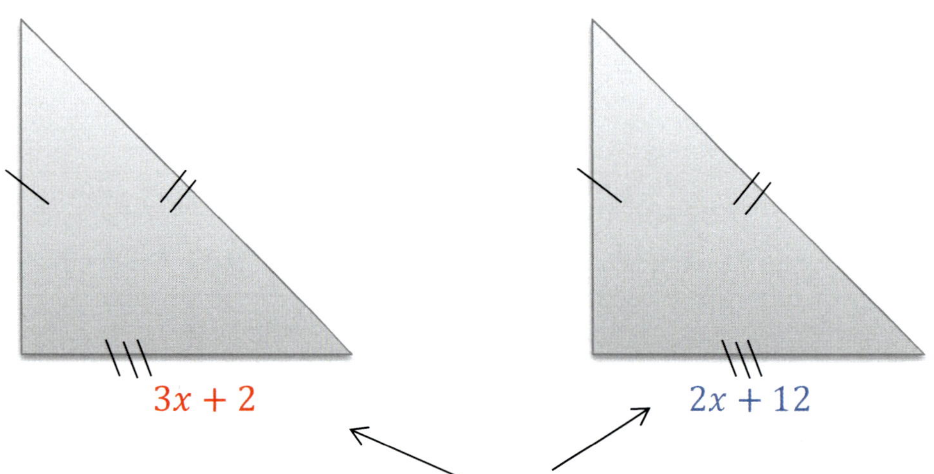

Do you know what to do? Since these sides are congruent, they are equal. I'll write that equation and then solve for x.

$$3x + 2 = 2x + 12$$

$$3x = 2x + 12 - 2$$
$$3x - 2x = 12 - 2$$
$$x = 10$$

Let's try another one. Solve for x in the equilateral triangle below.

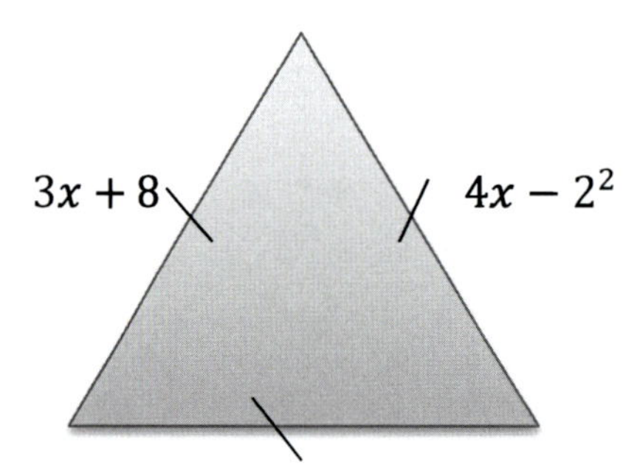

An equilateral triangle has three equal sides and the marks in the drawing above also prove that all three sides are equal. That means that $3x + 8 = 4x - 2^2$. Let's solve for x.

$$3x + 8 = 4x - 2^2$$

$$3x + 8 = 4x - 4$$

$$3x - 4x = -8 - 4$$

$$-x = -12$$

$$x = 12$$

Practice using algebra and geometry together by completing the next worksheet.

WORKSHEET 4

Name ____Date ________________

1. Below are two congruent triangles. Draw little lines to show which sides are congruent. Fill in the blanks with the appropriate values and then solve for x, y, and m.

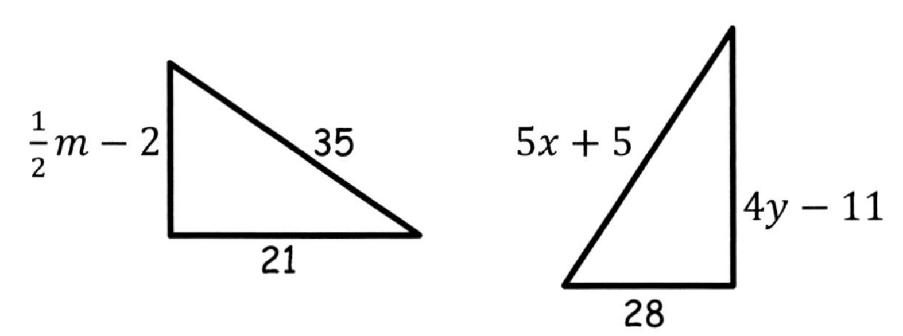

Side 1: ________ ≅ ___________

Side 2: ________ ≅ ___________

Side 3: ________ ≅ ___________

2. Look at the picture below. Solve for x.

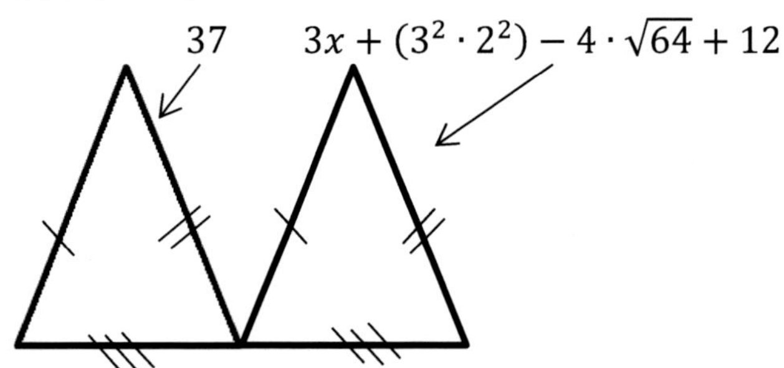

Worksheet 4 page 2

3. If $x = 4$, are the two triangles below congruent?

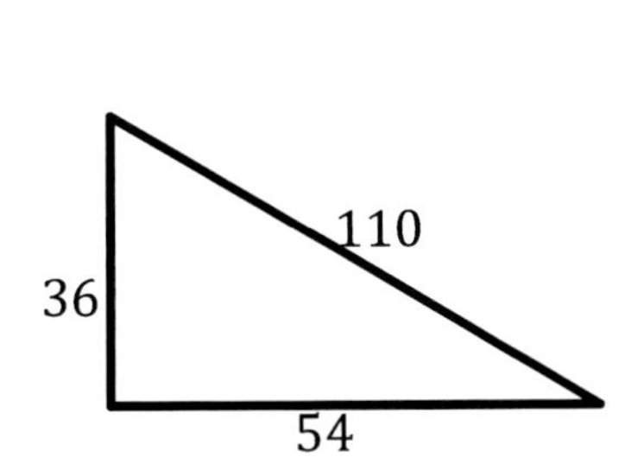

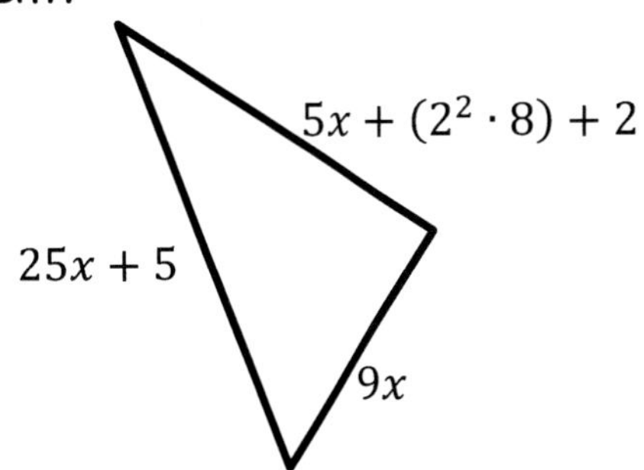

4. Every triangle has 3 angles. When you add all 3 angles together, it will always equal 180 degrees. Solve for x with the information given below.

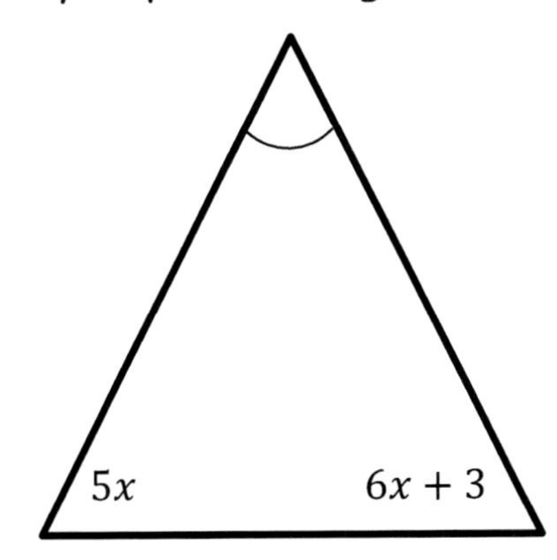

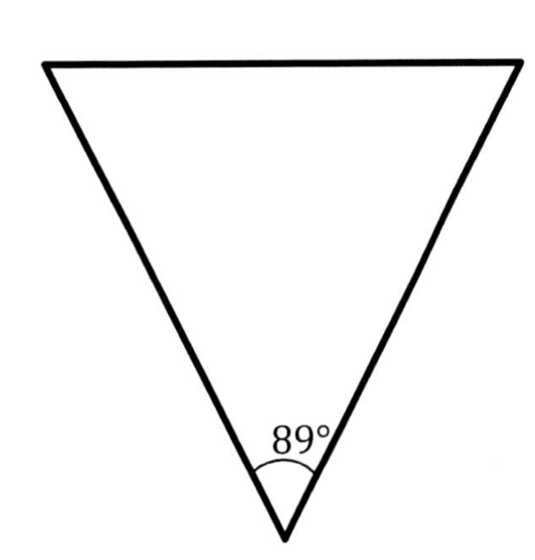

LESSON 5: SIMILAR TRIANGLES

In the last lesson, you learned that congruent triangles are exactly the same. They are the same size and the same shape with the same angles. Now you will learn about *Similar Triangles.*

Similar Triangles are...well...similar, but not the same size. The two triangles below are the same shape with the same angles, but the sides are different lengths. That makes them similar, but not congruent.

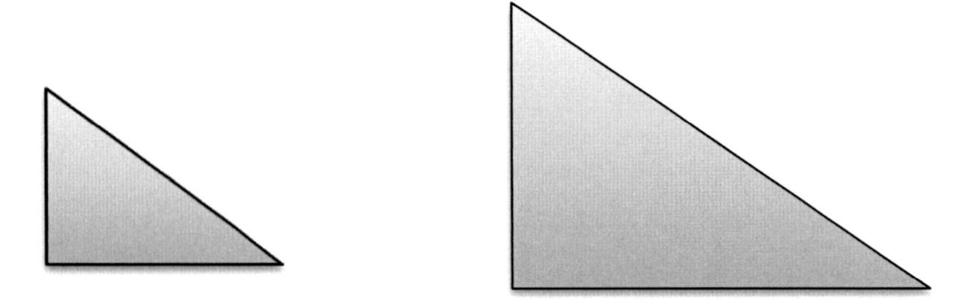

I realize this sounds ridiculous and you are probably wondering why I would even bother telling you this, but there is a point, I promise. First, we will learn about proportions because they are fun and then we'll learn how to use similar triangles in story problems, which is not as fun.

I have given a length for each side of the two similar triangles below. Now we will apply our proportions skills to see something cool.

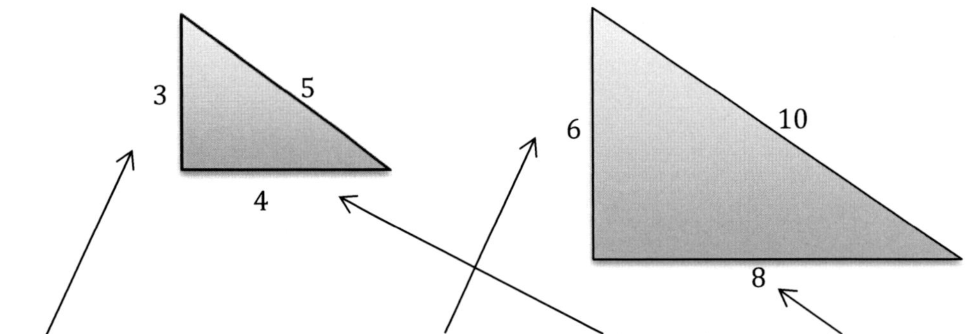

This side is (in proportion) to this side, as this side is to this side. I'll write that as a proportion.

$$\frac{3}{6} = \frac{4}{8}$$

This ratio means "3:6" (3 to 6) and this ratio means "4:8" (4 to 8). When you have "this ratio equals that ratio" you have a proportion. To prove that these two are equal, we can cross multiply. If the answer is the same in both directions, then they are equal.

$$3 \times 8 = \mathbf{24} \qquad 6 \times 4 = \mathbf{24}$$

$$\frac{3}{6} = \frac{4}{8}$$

Yep, they are equal. Now, let's suppose we didn't know one of those numbers. I will replace the 4 above with an x.

$$\frac{3}{6} = \frac{x}{8}$$

Now I'll cross multiply again, $3 \times 8 = \mathbf{24}\ and\ 6 \cdot x = \mathbf{6x}$.

$$24 = 6x$$

Solve for x.

$$4 = x$$

Tada! OK, that was easy. Take a look at these next two similar triangles.

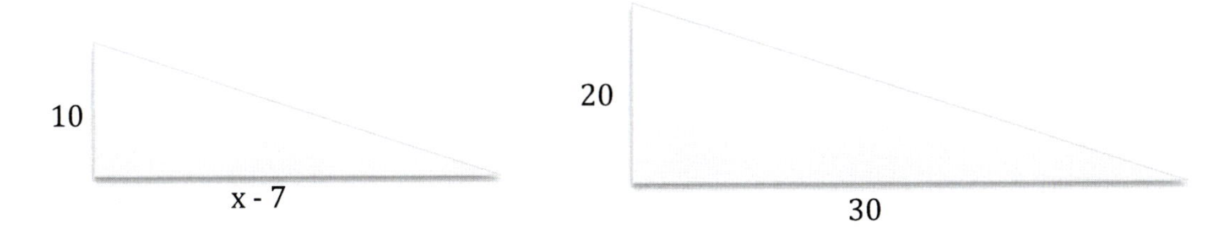

Let's do the same thing. I've written the proportions below.

$$\frac{10}{20} = \frac{x - 7}{30}$$

Again, we cross multiply and solve for x.

$$10 \cdot 30 = 300 \quad and \quad 20(x-7) = 20x - 140$$

$$300 = 20x - 140$$

$$300 + 140 = 20x$$

$$440 = 20x$$

$$22 = x$$

If x = 22, then the base of that last triangle is 15. But watch this! I'll bring back that last problem. It is also true to say that…

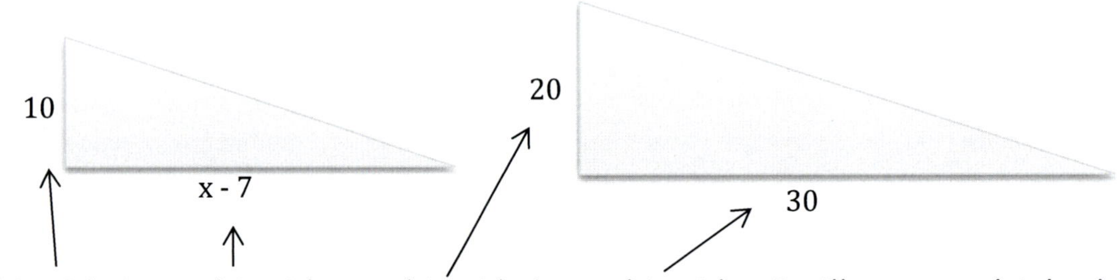

This side is to this side, as this side is to this side. I will cross multiply those ratios and then watch; I will get the same answer.

$$\frac{10}{x-7} = \frac{20}{30}$$

$$300 = 20x - 140$$

$$440 = 20x$$

$$x = 22$$

The trick is making sure each ratio is built the same way. In the first example, I was sticking to this formula:

$$\frac{Little\ Triangle's\ height}{Big\ Triangle's\ height} = \frac{Little\ Triangle's\ Base}{Big\ Triangle's\ Base}$$

The second time I solved the same example I used this formula:

$$\frac{Little\ Triangle's\ Height}{Little\ Triangle's\ Base} = \frac{Big\ Triangle's\ Height}{Big\ Triangle's\ Base}$$

Make sure your ratios are consistent and logical.

Next, I am going to overlap two triangles. Do you see the two similar triangles below? The big triangle is outlined in blue and the smaller triangle is outlined in black.

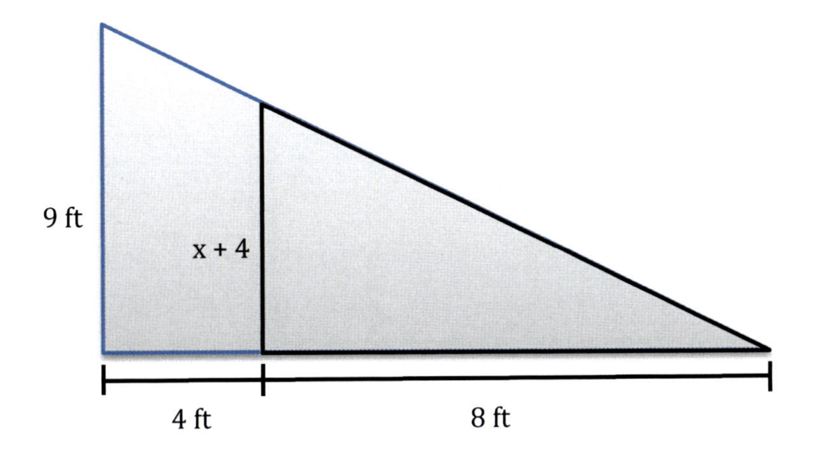

We are going to solve for x in the triangle above. It is solved the same way as before; just make sure the base of the big triangle is 12 ft.

Height → $\frac{9}{x+4} = \frac{12}{8}$ ← Base

Do you know what to do now? That's right, cross multiply.

$$\frac{9}{x+4} = \frac{12}{8}$$

$$72 = 12x + 48$$

$$72 - 48 = 12x$$

$$24 = 12x$$

$$2 = x$$

Now let's apply this knowledge to a story problem. Here is a classic story problem you might find on a math test.

A 6-foot-tall man is standing 24 feet from a light post. The length of his shadow created by the light is 4 feet. Find the height of the street light.

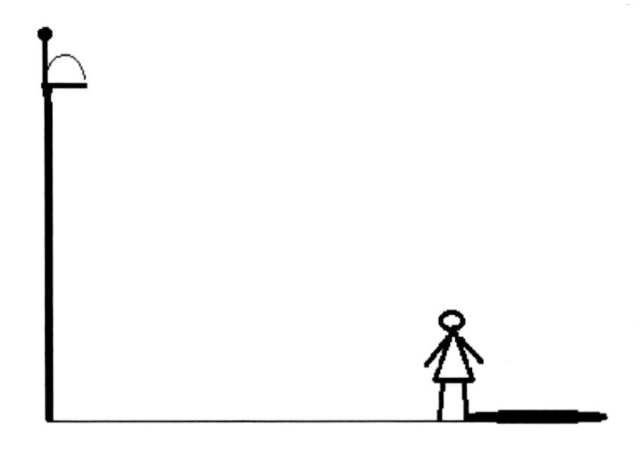

Do you know how to solve this problem? Believe it or not this is a Similar Triangle problem. To help us understand the question, I have added the measurements to the picture below.

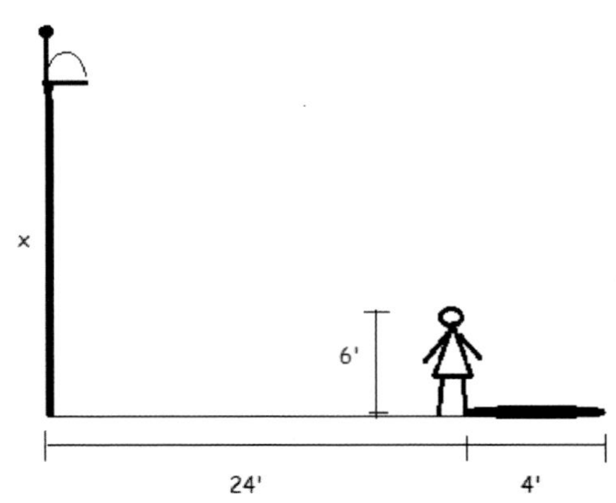

Next, I will draw the two similar triangles, over top of the picture.

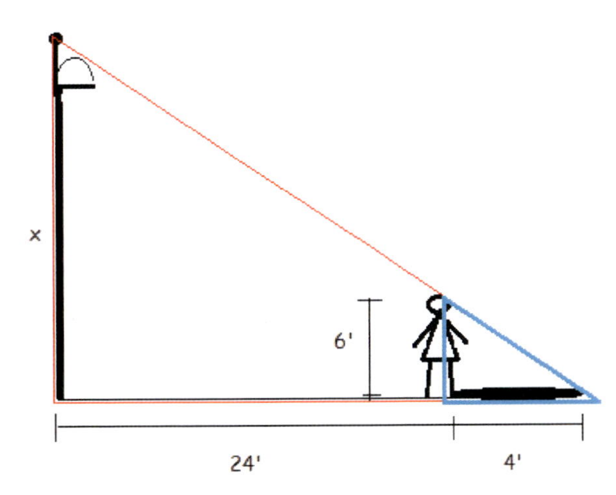

Now we can create an equation to help us solve for x. The larger triangle is "x feet" tall and the man is 6 feet tall, so the first ratio is $\frac{x}{6}$. The base of the big triangle is 28' and the smaller triangle's is 4'.

$$\frac{x}{6} = \frac{28}{4}$$

I will cross multiply to solve for x.

$$4x = 168$$

$$x = 42\ feet$$

The light post is 42 feet tall.

Home builders use similar triangles to describe how steep a roof is or will be. Just like a *slope* measures how steep a line is, builders use "pitch" to show how steep a roof is to be built. For example, let's say the pitch of a roof is 6:12. That means that for every 12 horizontal inches (or 12 feet), the roof will go up by 6 vertical inches (or feet).

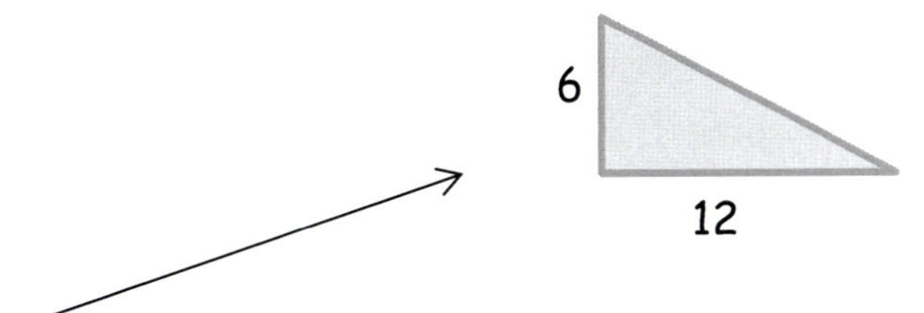

This shape is similar to the pitch of a roof, just much smaller. We can use it proportionately to find some unknown dimensions. Look at the roof on the next page. Half of the triangular rooftop is similar to our "6:12 pitch" triangle. So, if I know the length of the base or the height of the triangle, I could figure out the 3rd dimension.

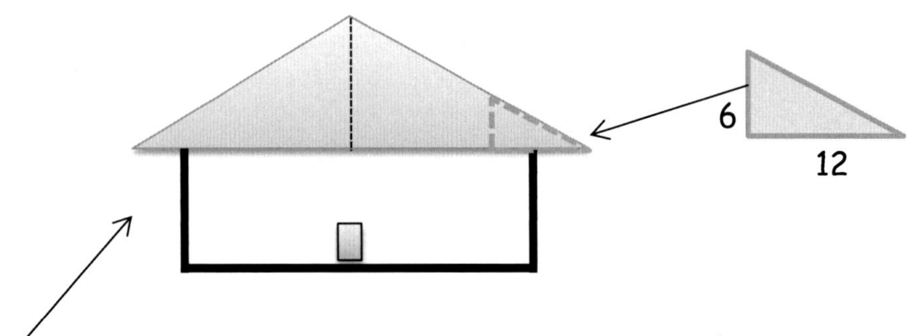

If the base of this triangle measures 36 feet, then half of it is 18 feet. Here are the ratios of our 2 triangles.

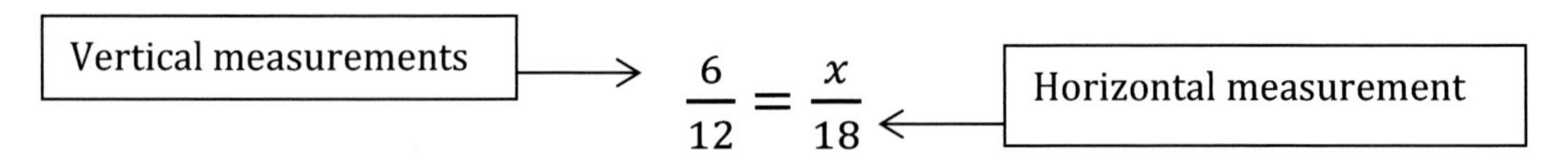

Cross multiply to solve for x.

$$12x = 108$$

$$x = 9$$

The height is 9 feet.

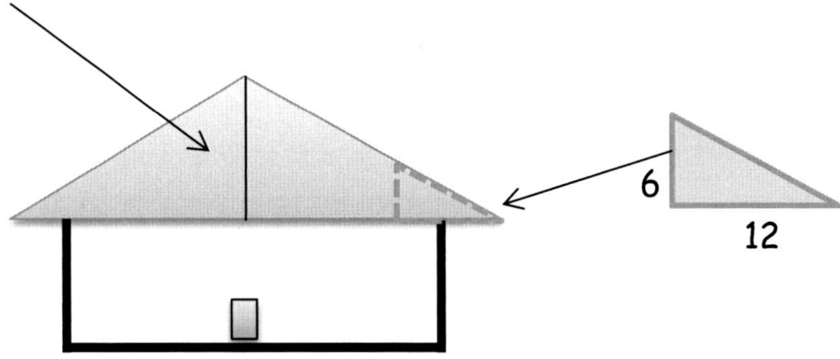

Just make sure that you are always consistent when creating proportions! By that I mean, the numbers on top must come from the same place. For example, let's say the best ratio of ice cubes:pop is 4 ice cubes per 12 ounces of pop. Less ice wouldn't make it cold enough and more ice would make the cup overflow. That ratio would be written like this:

$$\frac{4\ ice\ cubes}{12\ ounces\ of\ pop}$$

Then, if you wanted to find the amount of ice cubes needed for 48 ounces of pop, you would write this next ratio. JUST MAKE SURE THE TWO RATIOS BOTH HAVE ICE CUBES IN THE SAME SPOT.

$$\frac{4\ ice\ cubes}{12\ ounces\ of\ pop} = \frac{x\ ice\ cubes}{48\ ounces\ of\ pop}$$

It's OK if you switch around the ice cubes and the ounces, you'll get the same answer. Just make sure you stay consistent!

$$\frac{12\ ounces\ of\ pop}{4\ ice\ cubes} = \frac{ounces\ of\ pop}{x\ ice\ cubes}$$

Complete the next worksheet.

WORKSHEET 5

Name ____ Date ___________

Solve for "x" in the following sets of similar Triangles.

1.

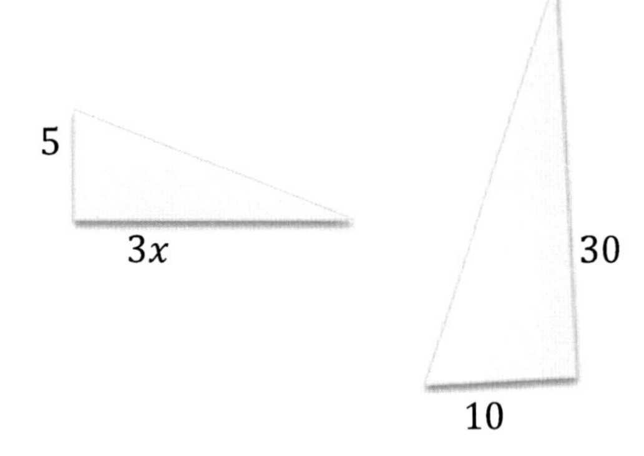

2.

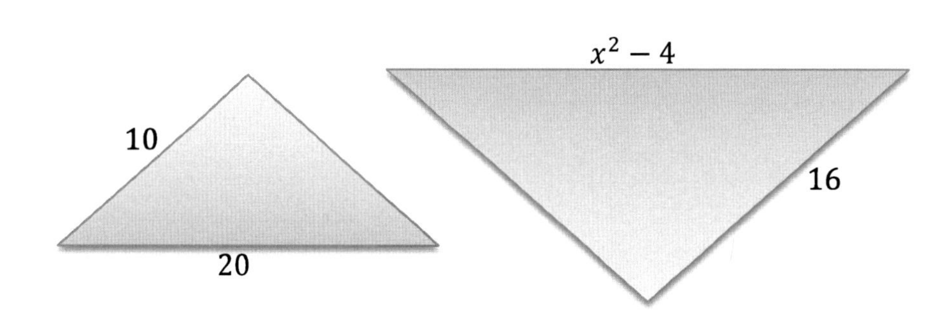

3.

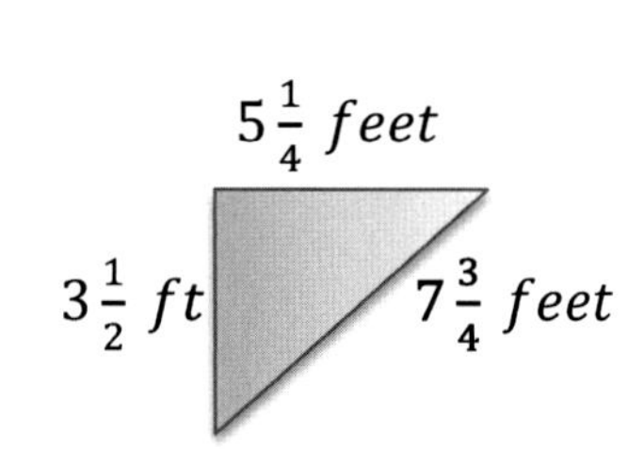

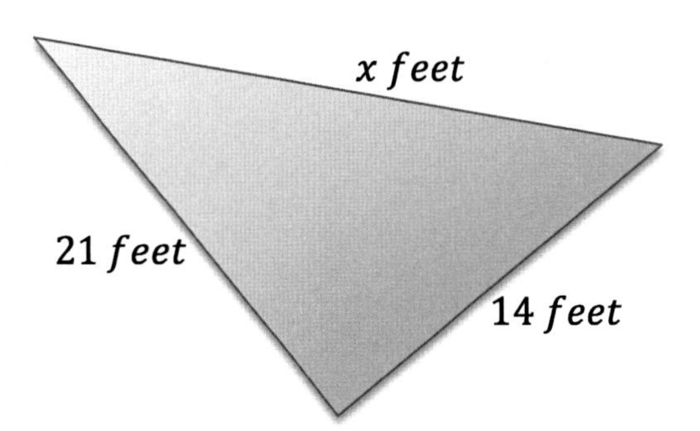

Worksheet 5 page 2

4. Solve for x in these two similar, isosceles Triangles.

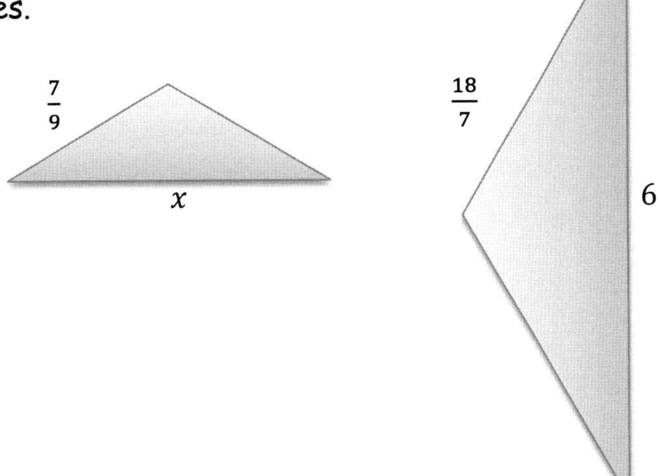

5. A builder wants to know the square footage of the shaded triangular area of the building below. His ladder isn't tall enough to measure the height of the triangle, so he will have to use math instead. He knows that the pitch of the roof is 5:12. Use similar triangles to find the area of the shaded area below. (HINT: Slice the gray triangle in half, so it is similar to the pitch).

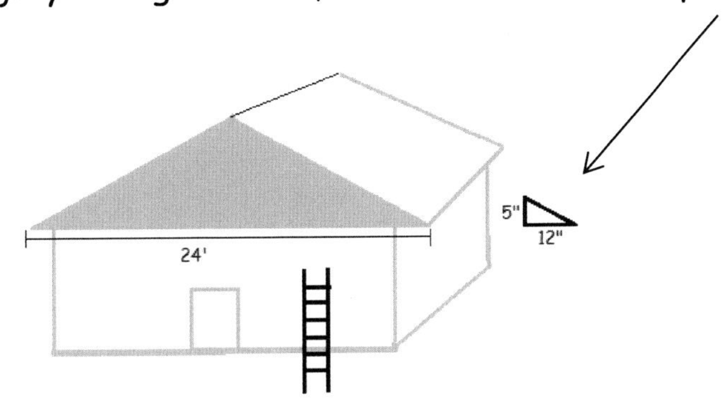

CHAPTER 1 TEST

Name ____ Date ____________

1. Judy needs to receive an average score of 94 in order to earn a scholarship. On the first test she received a score of 95. On the second test, she earned a score of 93. The third test was her worst score, it was a 92. What score must she get on her fourth test in order to earn the scholarship?

2. Daryl wants to save up $2000 to get his truck painted. He already has $750 saved up and he can earn $40.00 per hour at his job. How many hours does he need to work to earn enough money for the new paint job?

3. A horse is tethered to a pole in the center of a pen. The horse walks in a circle, stretching the rope as far as possible as he walks around the pole. Each time he circles the pole he travels 113.04 feet. How long is the rope?

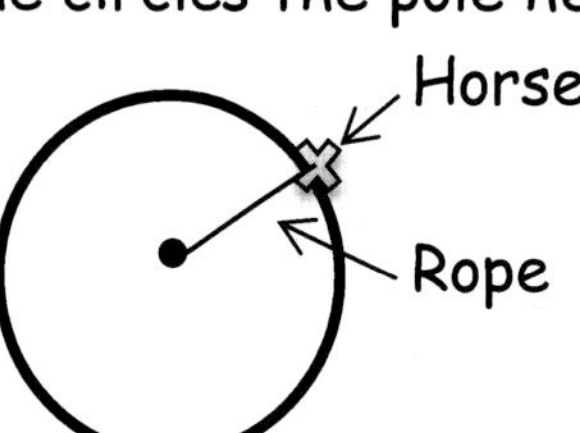

Chapter 1 Test page 2

4. A homemade rocket shot straight up into the air. The radar detector showed that it traveled at 110 mph. The stop watch showed that it traveled upwards for 1 minute before it started to fall. Approximately how far did the rocket travel?

5. A train is traveling east at 100 mph. Another train is heading straight towards it at 80 mph, 55 miles away. The trains will hit each other at the halfway point, if the tracks are not switched in time. The engineer can't switch the tracks for 15 minutes. Will the trains hit each other before the engineer can flip the switch?

6. Are the two isosceles triangles below congruent?

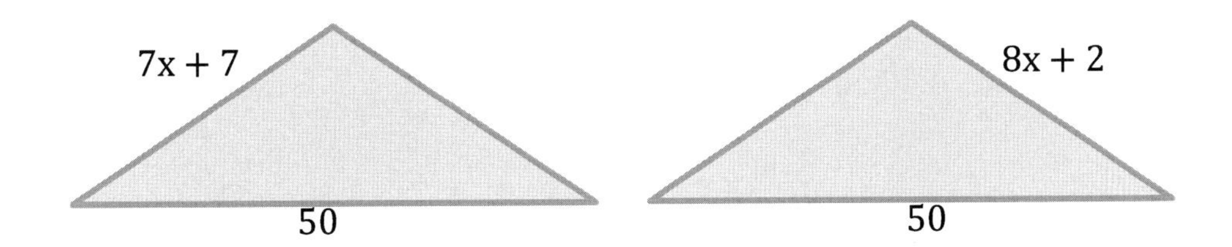

Chapter Test page 3

7. A builder needs to replace all of the blue boards on the house below. He charges $10 per foot, to replace the boards. How much will the builder charge?

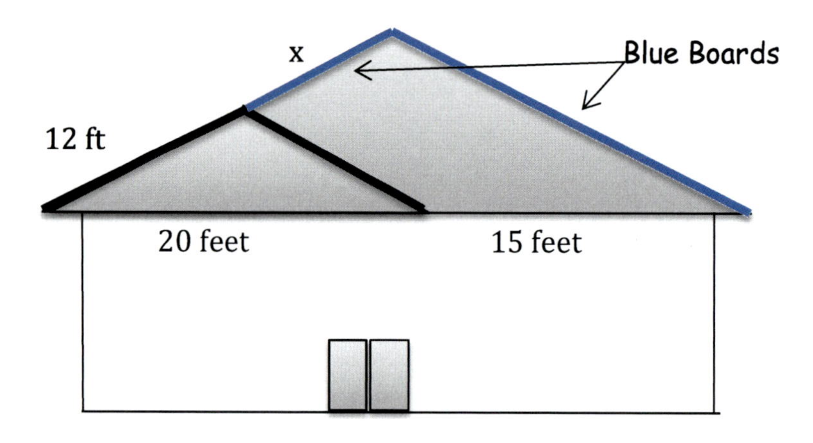

8. A factory can produce 2 guitars every 5 hours during the day shift and 4 guitars in 5 hours during the night shift. If each shift works 12 hours per day, how many days will it take to fill an order of 72 guitars?

Chapter Test page 4

9. Jo-Jo is going to drive from Tacoma to Federal Way, drop off an envelope and then drive back to Tacoma. This trip will take 40 minutes to complete. If he continues at the same rate of speed, how long will it take him to drive to Kent?

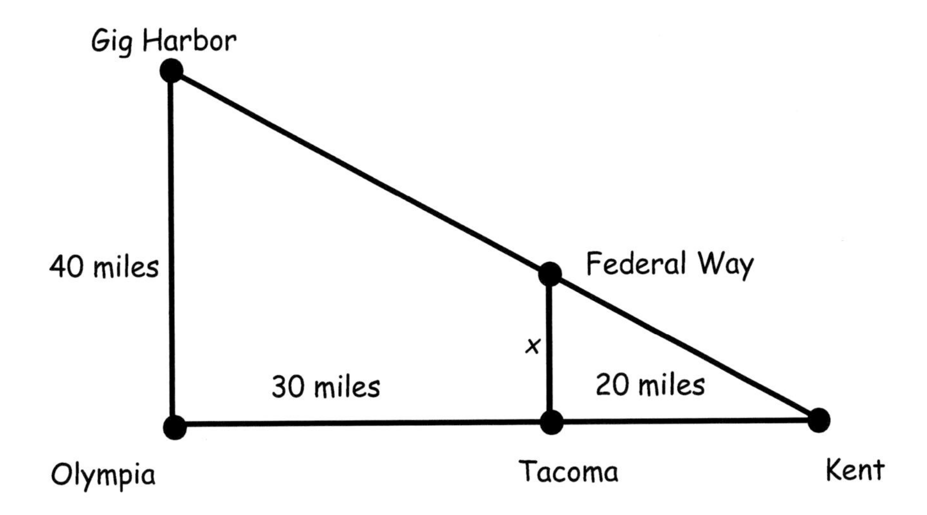

CHAPTER 2 PROBABILITY, INEQUALITIES, SOLUTION, AND MIXTURE PROBLEMS

LESSON 6: PROBABILITY

If you grabbed an apple from out of a bag that held 10 green apples and 1 red apple, you would PROBABLY get a green apple. It isn't 100% certain that you will get a green one because there is a small chance you will pull out the only red apple, but you will PROBABLY get a green one.

That's what *Probability* is all about. Trying to find out how likely it is that something will happen. By using some "Probability Skills" we can come up with a number between 0 and 1 to show the probability of something happening. The number 1 means it is certain to happen and 0 means it will never happen. All the numbers in between will show the probability of something happening.

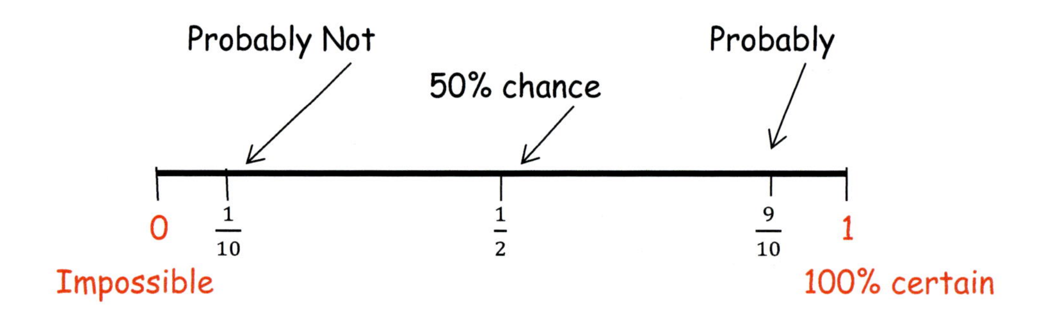

For example, if you flip a coin, how likely is it to land on "heads?"

It will either be heads or tails; those are the only two possibilities. So, let's say you flip a coin one time. The outcome will be one of two possibilities.

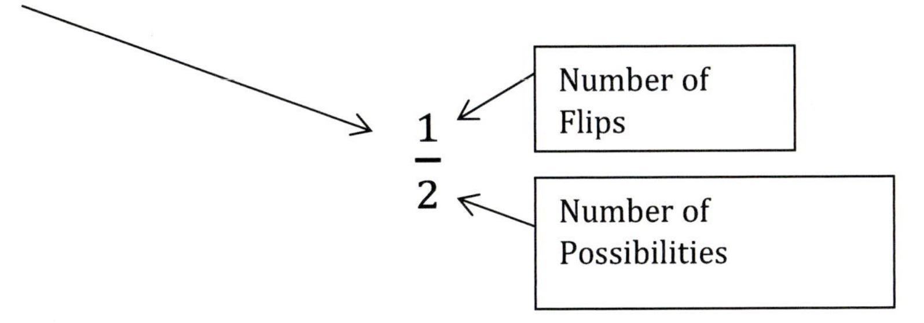

That's all there is to it. The probability of a flipped coin landing on "heads" is 1 in 2. It is written like a fraction. It kind of looks like one half, doesn't it? That's because it is! The coin is expected to land on "heads" half of the time.

Now, think about a deck of cards. There are 52 cards in a full deck, but there is only one 9 of clubs in the deck. The probability of randomly pulling the 9 of clubs out of the deck is 1 in 52, or $\frac{1}{52}$.

Do you know the probability of pulling a 3 from the deck of cards? There are four different 3's: the 3 of hearts, 3 of clubs, 3 of spades, and the 3 of diamonds. What is the probability of pulling a 3 out of a full deck of cards?

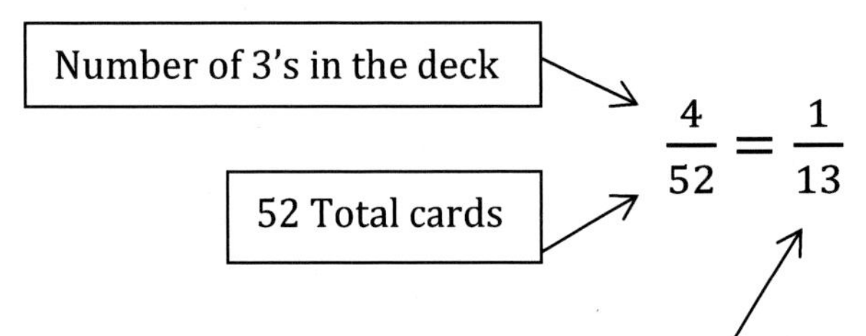

Probabilities should always be reduced down as far as possible or written as a percentage or decimal number.

That seems easy enough, doesn't it? All you have to do is fill in the formula below with your numbers and you've got it. I will give you a few more examples of probabilities. As you read them, try to fill in the formula on your own before reading my solutions.

$$Probability\ of\ an\ event\ happening = \frac{Number\ of\ ways\ it\ can\ happen}{Total\ number\ of\ outcomes}$$

OR

$$p = \frac{\#\ of\ ways\ possible}{Total\ \#\ of\ outcomes}$$

A box holds a dozen eggs, 3 of them are broken. What is the probability of selecting an egg that is NOT broken?

$$probability = \frac{Number\ of\ unbroken\ eggs}{The\ whole\ dozen} \qquad p = \frac{9}{12} = \frac{3}{4}$$

The probability of not selecting a broken egg is $\frac{3}{4}$. This is super easy, isn't it? It gets a little more difficult, but not much.

A bag of 24 potatoes has 4 rotten potatoes. What is the probability that you will select a rotten potato from the bag?

$$probability = \frac{Rotten\ potatoes}{All\ the\ potatoes} \qquad p = \frac{4}{24} = \frac{1}{6}$$

The probability of selecting a rotten potato is $\frac{1}{6}$.

In a bag of 36 marbles there are 10 blue marbles, 6 red marbles, 10 green marbles, and 10 multi-colored marbles. What is the probability of selecting a red marble from the bag?

$$Probability = \frac{Red\ marbles}{All\ 36\ marbles} \qquad p = \frac{6}{36} = \frac{1}{6}$$

Nearly every higher math test will have at least a few probability questions. The ones we've done together were easy, so let's take a look at some more challenging problems.

If a die is tossed twice, what is the probability of getting a 3 on both tosses?

This one is a little different, but just as easy. First of all, what is the probability of getting a 3 when you toss the die once? It is 1 in 6, right?

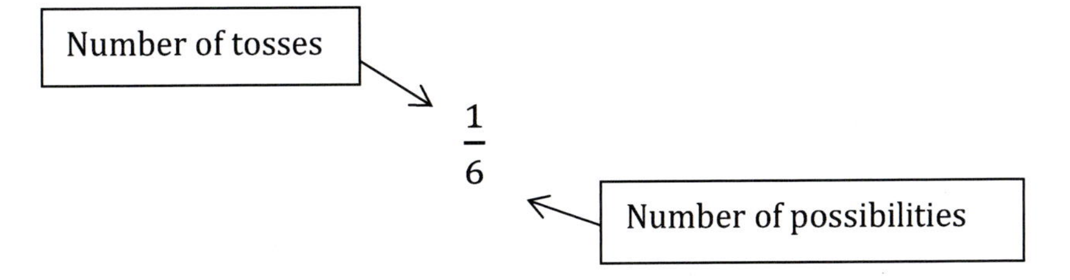

And what is the probability of rolling a 3 the second time? It is the same number.

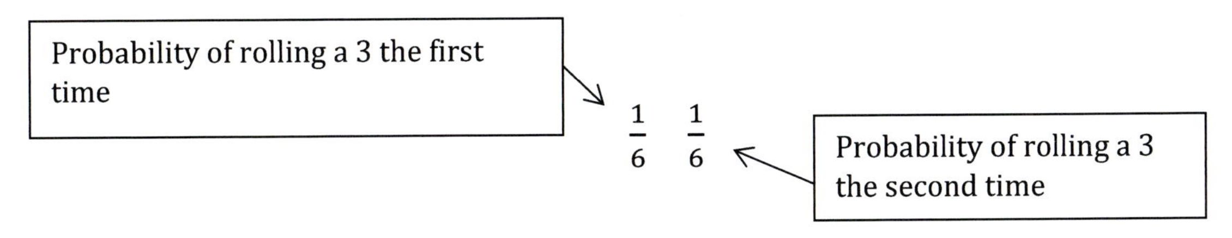

Now comes the hard part. Multiply them together. Yep, that's it, just multiply them together.

$$\frac{1}{6} \cdot \frac{1}{6} = \frac{1}{36}$$

I'm not exactly sure how or why this works. But go ahead and roll a die twice. Count how many times it takes before you get it to land on 3, both times. I bet it will take about 36 tries or it will take 72 tries to get it to happen twice. It's not an exact science, but it is pretty close.

When you roll a die, there are always 6 possible outcomes. Whether it is the first roll, the second, or the third, there will always be 6 possibilities. But not all probability questions are like that, as you will see in the next example.

A test shows that 10 out of 100 toy helicopters are defective. In a random sample, two helicopters are chosen. What is the probability that both are defective?

This one is a little different because once you take out the first helicopter, the total number of helicopters changes. When we select our first helicopter, the probability of it being defective is 10 in 100.

$$\frac{10}{100}$$

But when we select the second sample, there are only 99 helicopters left and since it is possible that we took one of the defective toys the first time, the number on top changes too. The probability of picking a defective helicopter on the second try is 9 in 99.

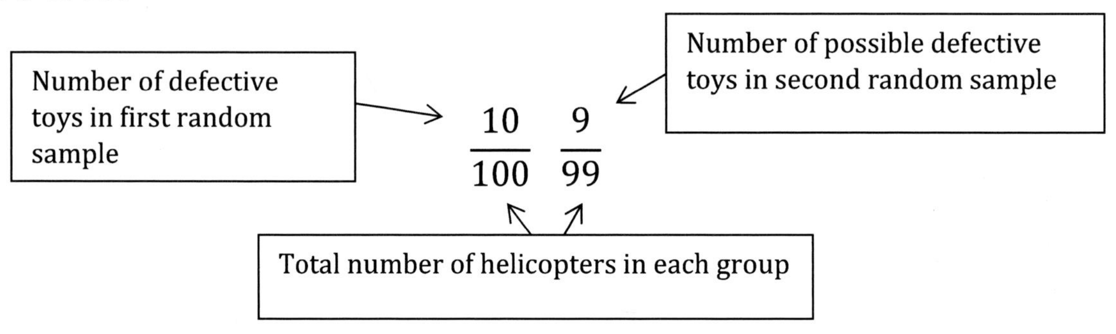

All we have to do now is multiply the two fractions together and reduce.

$$\frac{10}{100} \cdot \frac{9}{99} = \frac{90}{9900} = \frac{9}{990} = \frac{1}{110}$$

There you have it. The probability of selecting two defective helicopters is 1 in 110. Not very likely, is it?

Let's try another one together.

Two cards are drawn from an ordinary deck. What is the probability that both cards have different suits?

This one takes a little bit of thinking. The first card you draw can be ANY card, it doesn't have to be just a 3 or just an ace. It can be any card. So, what is the probability of drawing a card from a deck of cards? Well, it's pretty good. In fact, there are 52 different possibilities, so that would be $\frac{52}{52}$, which equals 1.

Do you remember the number line I drew a few pages back? It went from 0 to 1. The number 1 meant 100% certain. The probability of drawing a card from a deck of cards is $\frac{52}{52}$, so it is 100% certain that we will draw a card, get it?

Now when we draw the next card, there will only be 51 cards left, right?

$$\frac{52}{52} \quad \frac{}{51}$$

Do you know how many suits there are in a deck of cards?

There are 4 different suits; each one has 13 different cards. The question asks for the probability of the two cards having different suits. Let's assume we drew a "heart" the first time.

How many "hearts" are left in the deck now? There are 12 "hearts" left. All the other cards have different suits. So how many cards are left that have a suit besides hearts?

$$51\ cards - 12\ hearts = 39\ non\ hearts$$

So if 39 of the 51 cards are not "hearts," then the probability of drawing two cards with different suits is:

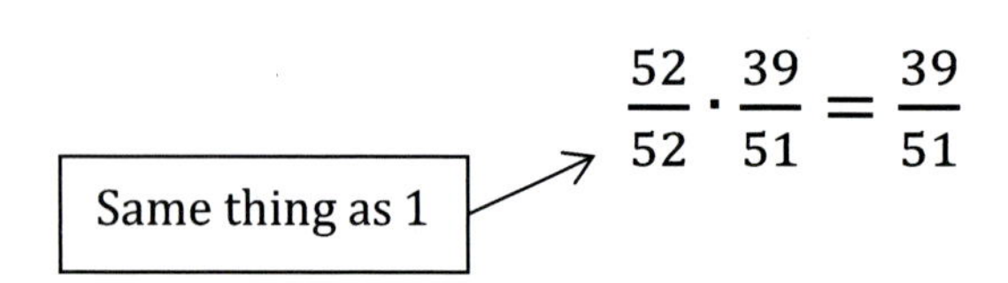

$$\frac{52}{52} \cdot \frac{39}{51} = \frac{39}{51}$$

Those last three examples all have something in common; the word "and." The first example wanted to know the probability of tossing a 3 **AND** then another 3. The second example wanted to know the probability of selecting a defective helicopter **AND** then another defective helicopter. The third example wanted the probability of drawing a card **AND** then another card of a different suit.

Since all three problems used the word "**and**" I had you multiply the two probabilities together. We will call that the "And Rule."

Next, we will solve an "**OR**" question, instead of an "**and**" question. I'll explain what I mean with the next example.

What is the probability of obtaining all heads ***OR*** *all tails when three coins are tossed?*

Since this question uses the word "OR," we have to do some **adding** too. We will call this the "Or Rule."

So, let's take a look at our problem. It wants to know the probability of tossing a coin three times and having it land on heads all 3 times **OR** tails all 3 times. Let's focus on the first half of the question: What is the probability of tossing a coin three times and having it land on heads all three times?

The first toss has a 1 in 2 chance of landing on heads.

$$\frac{1}{2}$$

The second toss also has a 1 in 2 chance of landing on heads.

$$\frac{1}{2} \quad \frac{1}{2}$$

And so does the third.

$$\frac{1}{2} \quad \frac{1}{2} \quad \frac{1}{2}$$

Let's multiply these three fractions together, to get the probability of the coin landing on heads all three times.

$$\frac{1}{2} \cdot \frac{1}{2} \cdot \frac{1}{2} = \frac{1}{8}$$

NOW we will get to the "OR" portion of our story problem. The question says "or tails" so we need to ADD that probability to the first.

What is the probability of the coin landing on "tails" all three times?

$$\frac{1}{2} \cdot \frac{1}{2} \cdot \frac{1}{2} = \frac{1}{8}$$

It's the same as heads! Now that we know the probability for both, we ADD them together.

$$\frac{1}{8} + \frac{1}{8} = \frac{2}{8} = \frac{1}{4}$$

The probability of tossing a coin three times and it landing on heads each time or on tails is 1 in 4.

Practice finding probabilities on the next worksheet.

WORKSHEET 6

Name ______________________________________ Date ______________

1. In a box of 100 animal cookies, there are 4 different animal shapes. There are 25 bears, 20 giraffes, 30 lions, and 25 camels. What is the probability of randomly selecting a lion shaped cookie out of the box?

2. I have written down some names and put them in a hat. Six of them are girl names and seven of them are boy names. If I draw a name out of the hat, what is the probability of it being a boy's name?

3. There are a dozen light bulbs in a box. Three of them are burnt out. What is the probability of selecting two light bulbs that both work?

4. I roll two dice at the same time. What is the probability of them both landing on 5?

5. Two cards are selected from an ordinary deck of cards. What is the probability of them both being black or both being red?

LESSON 7: INEQUALITIES

Do you recall learning about the "Less Than" and "Greater Than" signs?

$$< \quad >$$

Sometimes these signs are used in algebra to show *Inequalities*. Take a close look at that word, it means *not equal.*

In - equal - ities

Here is an example of an Inequality:

$$x + 3 < 12$$

I can't call this an *equation* because an equation implies that the two sides are equal, which they are not. The Inequality above states that $x + 3$ is *less than* 12, not equal to 12. An Inequality is solved pretty much the same way as an equation, except for one little twist. First, I'll show you how they are the same and then I'll show you the one little twist.

Inequality	Equation
$x + 3 < 12$	$x + 3 = 12$
$x < 12 - 3$	$x = 12 - 3$
$x < 9$	$x = 9$

Read through the math above. Do you see how they are solved the same way? Most SAT tests or College Placement Tests will have at least one of these inequalities for you to solve. They can be solved the same way as an algebraic equation UNTIL you multiply or divide by a negative number. Once you do that, you have to flip the "less than" or "greater than" sign around. I'll show you what I mean with some easy math.

$$5 < 8$$

You know this is true, right? Five is less than eight. I can add 2 to both sides, and it will still be true.

$$2 + 5 < 8 + 2$$
$$7 < 10$$

I can multiply both sides by 3 and it will still be true.

$$3 \cdot 5 < 8 \cdot 3$$
$$15 < 24$$

But watch what happens when I multiply both sides by a negative number.

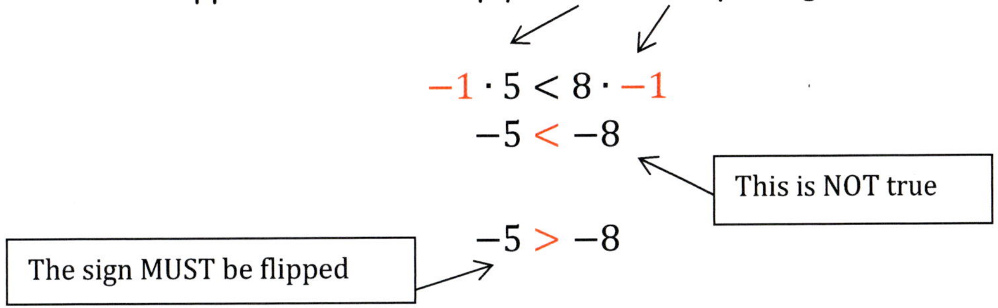

Do you see why I said that the inequality above is not true? Because negative 5 is NOT less than negative 8, it is greater than negative 8. So, the sign must be flipped, in order to be correct.

The same is true when you divide by a negative number; the sign must be flipped. Look at this next example. Four is less than six, right?

$$4 < 6$$

But if I divide both sides by negative 2, the sign must be flipped.

$$\frac{4}{-2} < \frac{6}{-2}$$

$$-2 < -3$$

NOT TRUE
FLIP THE SIGN

$$-2 > -3$$

The reason you need to know this is because sometimes you will be given an inequality like the one below.

$$-2x < 30$$

In order to get x by itself, we will need to divide by negative 2.

$$\frac{-2x}{-2} < \frac{30}{-2}$$

Once we divide by a negative number, what MUST we do? That's right, flip the sign around.

$$x > -15$$

Let's try a challenging one together, before you complete the worksheet on your own. Here is our next example.

$$\frac{2x-1}{3} < 1$$

This is solved the same way as an equation. We want x to be by itself. The first step is to multiply both sides by 3. Should we flip the "less than" sign around?

$$\frac{\overset{1}{\cancel{3}}}{1} \cdot \frac{2x-1}{\underset{1}{\cancel{3}}} < 1 \cdot 3$$

$$2x - 1 < 3$$

No, we don't flip it around because we didn't multiply or divide by a negative number. The next step is to add 1 to each side.

$$2x - 1 + 1 < 3 + 1$$

$$2x < 4$$

Do we flip the sign yet? No, not yet. Do you know what to do next? Divide both sides by 2, to get x by itself.

$$\frac{2x}{2} < \frac{4}{2}$$

$$x < 2$$

Do we flip the sign now? No, we never multiplied or divided by a negative number, so the sign remains the same.

Practice solving inequalities on the next worksheet.

WORKSHEET 7

Name ______________________________ Date ______________

Solve the Inequalities below.

1. $x - 8 > 5$

2. $-10x < 20$

3. $-2x > -12$

4. $5x - 4 < 2x + 8$

5. $\frac{3x-1}{2} < 10$

6. $4x - 3 < 2x + 7$

7. $\frac{3x+5}{2} < 7$

LESSON 8: SOLUTION PROBLEMS

In this lesson, you are going to learn about Solution Problems. When I say "solution," I don't mean the answer to a problem. I'm talking about a solution made of liquids. I think "potion" would have been a better description, but that's just me.

Let me give you a few examples of a solution. A restaurant owner might pour a little bleach into water to create a sanitizing solution. That's one type of a solution. Or a painter might mix paint thinner with paint to have a solution of paint that isn't so thick. That is another type. In math, it is common to see a solution problem that involves saltwater or acid and water.

Let's go over an example of a solution problem together. I will use a type of solution that everyone is familiar with; chocolate milk! Below, are two glasses with seven ounces of chocolate milk in each glass. The first glass has way too much chocolate syrup in it. It is 40% syrup and 60% milk. The second glass doesn't have near enough chocolate flavoring. It is only 5% syrup. If I mix the two glasses together, I would have 14 ounces of improved chocolate milk. What would the percentage of chocolate be in the new glass?

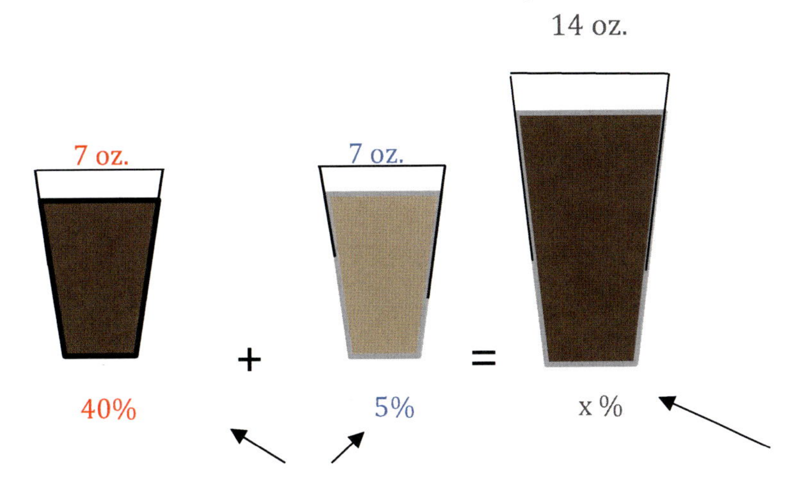

To solve this problem, we need to add the two small glasses together and make them equal the big glass. But we can't just add 7 ounces plus 7 ounces and get fourteen ounces. NO! That won't work. Each glass has a PERCENTAGE attached to it, so we MUST multiply the quantity of liquid (7 ounces) by the percentage of chocolate, first!

Look at the numbers, in red, on the first glass. The amount of liquid is 7 ounces. The percentage of chocolate is 40%. We need to multiply those two number together. Here is the math:

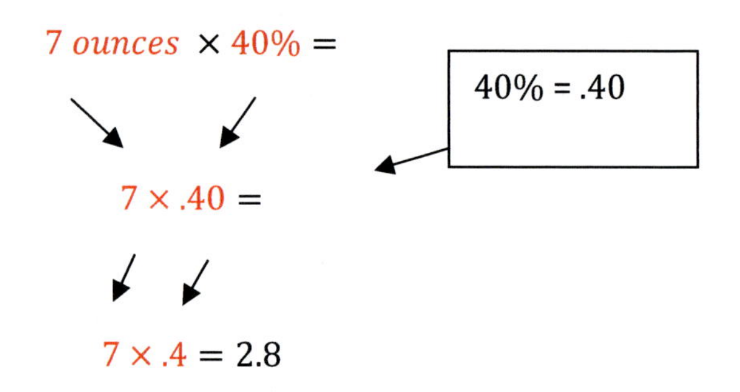

First glass with too much chocolate equals: 2.8

The numbers, in blue, on the second glass are 7 and 5%. I'll change that 5% to .05. I must keep the zero there, otherwise, it would equal 50%. And that is not right.

Second glass with too little chocolate $7 \times .05 = .35$

The "numbers" on the final glass are 14 and x%. I'll drop the percentage sign for now, but don't let me forget that "x" is a percentage, not a whole number. So, if I end up with x = .30, then the answer should be written $x = 30\%$. I'll write the entire equation and then we'll solve for x.

$$(7 \times .4) + (7 \times .05) = 14x$$

1st glass 2nd glass Big glass

$$2.8 + .35 = 14x$$

$$3.15 = 14x$$

$$\frac{3.15}{14} = x$$

$$.225 = x$$

Oh, that's right. Thanks for reminding me. We are looking for a percentage, so let's move the decimal point over two digits. Now $x = 22.5\%$. That is the percentage of chocolate syrup in our 14-ounce solution.

Do you get how that works? In a solution problem, each solution will have an amount and a percentage. You just multiply the two numbers from each container and then write an addition equation. Of course, one of the values will be missing because this is algebra. We call that missing number "x." Our equation helps us solve for x.

Sometimes, these problems are called "weighted problems" because each amount has a "weight," so to speak. In the last example, the first seven-ounce glass had a "weight" of .4 while the amount of chocolate in the second glass "weighed" less.

Let's try another one. I'll stick with chocolate syrup and milk, but I'll change the amounts. Here is the word problem:

How many ounces of 20% chocolate milk solution do you need to add to 40 ounces of 50% chocolate milk to end up with a solution that is 30% chocolate and milk?

Do you understand what that question is asking? In this problem, we have 40 ounces of chocolate milk with 50% chocolate in it. We need to add some amount - but we don't know how much - of 20% chocolate milk to it, until it is diluted down to 30% chocolate milk.

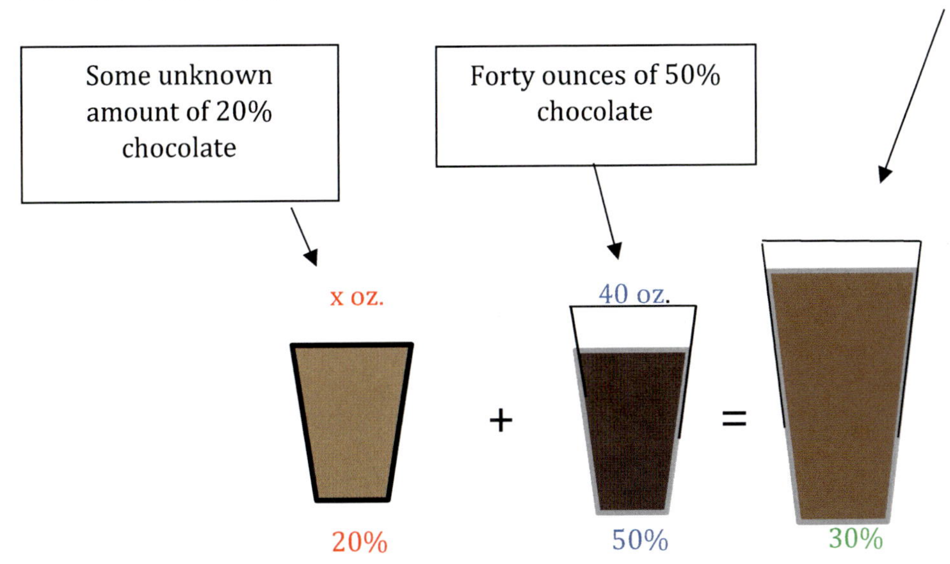

We don't know how much of the first solution we will use, so that amount is "x." So, how much will be in the big glass? For sure it will have 40 ounces in it, from the second glass, but how much will we pour in from the first glass? We don't know! That's why it is called "x." That means that the big glass will have "x + 40" ounces! Do you get that? That's "x" plus "40" equals "x + 40." Let's do the math.

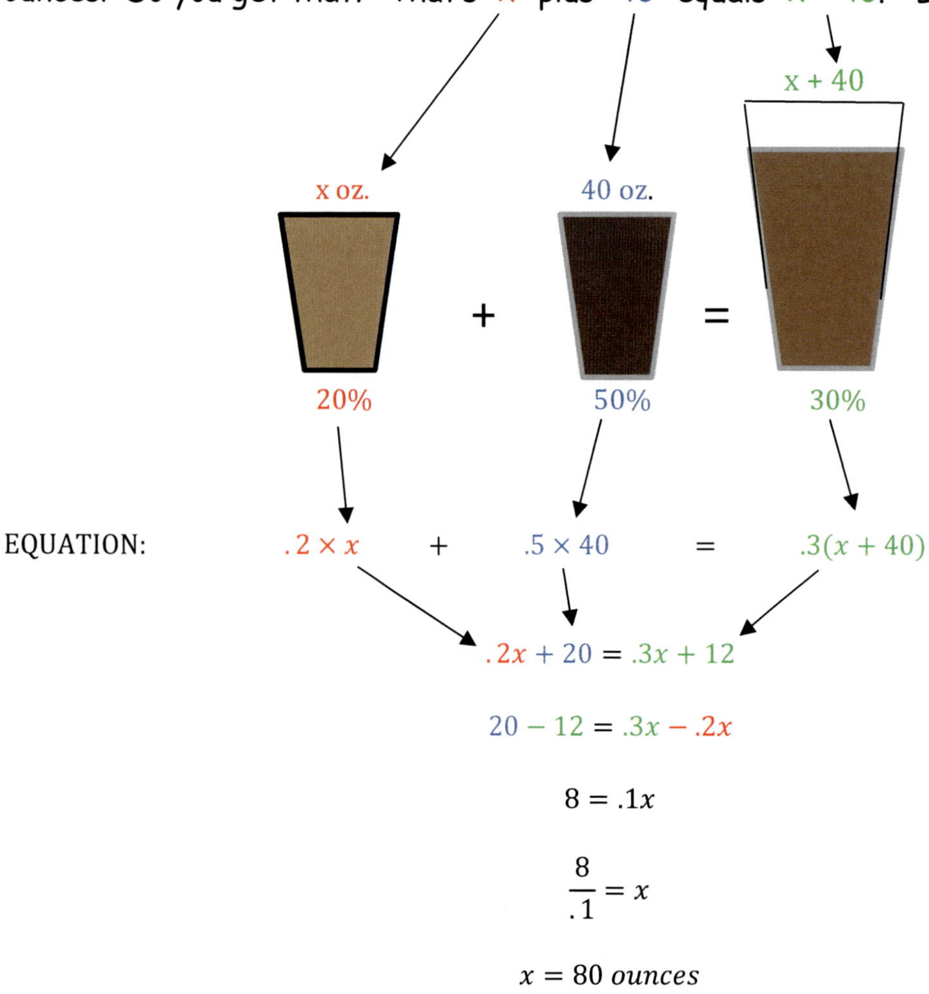

You would need to add 80 ounces of that 20% chocolate milk to the 50% chocolate milk to get a solution that is 30% chocolate.

Now let's try one that doesn't involve chocolate milk. This time we will use a solution of alcohol and water. We have 100 cups of this solution and it is 25% alcohol. It needs to be diluted with pure water until it is down to 10% alcohol. How much pure water should we add?

This one is a little different. What is the "weight" of alcohol in pure water? There is no alcohol in pure water, so the answer is zero. Here it is in picture form:

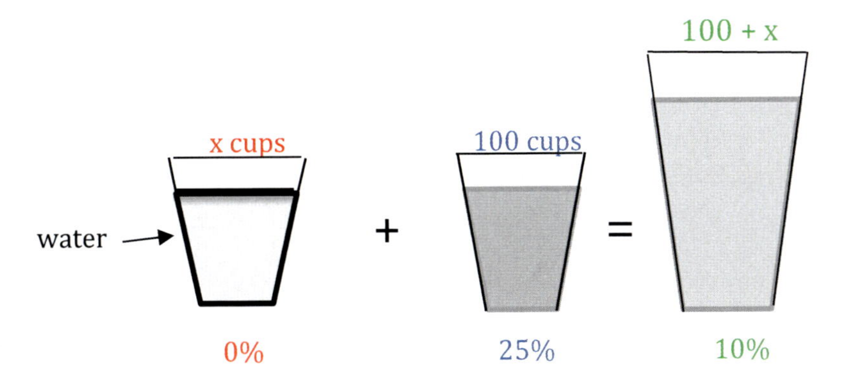

Let's build an equation. But wait! Look at the container of pure water. How much is x times zero? Well, anything times zero equals zero, so...I guess it's zero! Let's continue.

$$(0 \times x) + (.25 \times 100) = .1(100 + x)$$

$$0 + 25 = 10 + .1x$$

$$25 - 10 = .1x$$

$$15 = .1x$$

$$\frac{15}{.1} = x$$

$$x = 150\ cups$$

You would need to add 150 cups of pure water to the container of 25% alcohol to get a solution that is 10% alcohol.

Let's try one more together.

15 gallons of an alcohol and water solution is 12% alcohol. It is added to 30 gallons of a solution that is 45% alcohol and water. What is the percentage of alcohol in the new solution?

I think I can solve this one without using any pictures. I'll just out write the equation this time.

First Solution: $15\ gallons \times .12 = 1.8$

+

Second Solution: $30\ gallons \times .45 = 13.5$

=

Final solution: $45\ gallons \times x = 45x$

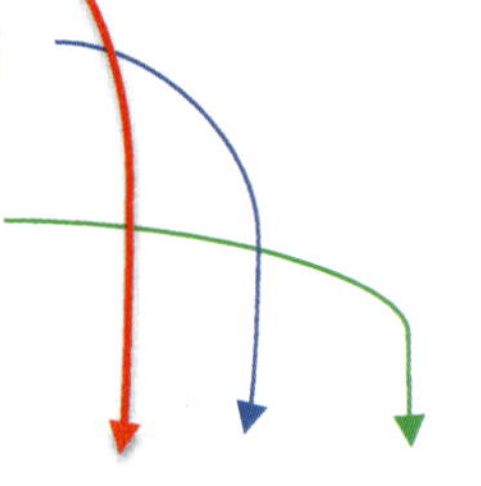

Equation: $1.8 + 13.5 = 45x$

Solve for x: $15.3 = 45x$

$.34 = x$

$x = 34\%$

Make sure you always go back to the original problem and read the questions again. My answer is x = 34%. The new solution would be 34% alcohol and water. If you think you've got it, complete the next worksheet. If you are a little uncertain, please read this lesson again.

Name ________________________________ Date ______________

WORKSHEET 8

1. A container holds 18 liters of a solution that is 7% acid. It is poured into 12 liters of a 20% acid and water solution. What is the percentage of acid in the new solution?

2. You have a container with 32 ounces of a solution of 23% alcohol and water. Your assistant accidentally pours 8 ounces of an alcohol and water solution and now your container is down to 20% alcohol. What was the percentage of alcohol in your assistant's glass?

Worksheet 8 page 2

3. How much of a 5% iodine and water solution should be added to 4 cups of 2% iodine and water to create a solution that is 3% iodine and water?

4. How much pure water should be added to 180 ml solution of bleach and water that is 23% bleach to dilute it down to a solution that is 10% bleach and water?

Worksheet 8 page 3

5. A gas can contains twelve cups of an oil and gasoline solution. It is currently 60% oil, but it needs to be down to 45%. How much pure gas should be added to the solution to get it to 45% oil?

6. If you mix 135 cm^3 of a solution that is 17% alcohol and water with 80 cm^3 of a 53% alcohol and water solution, what will the new percentage of alcohol?

LESSON 9: SOLUTION PROBLEMS WITH TWO VARIABLES

In the last lesson, you learned how to solve a solution problem that had one missing value; "x." In this lesson, you will learn how to solve a solution problem with two missing values; "x and y." The math looks very impressive and super complicated, but it's really kind of easy.

Here is our first example. We will go through it together.

How many pints of a 6% chlorine and water solution and a 12% chlorine and water solution should be combined to create 15 pints of a 9% chlorine and water solution?

In this problem, we have three different containers just like last time. The first container has a solution of 6% chlorine and water. The second container holds 12% chlorine and water, but we don't know how many pints are in either container. We have to pour a little bit of each solution into the third container until we have 15 pints of chlorine and water that is 9% chlorine. No problem, watch this:

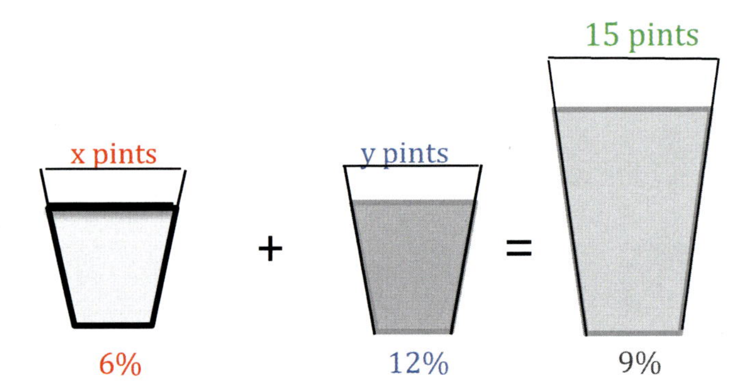

Focus your attention on the number of pints on top of each container, for a moment.

$$x\ pints + y\ pints = 15\ pints$$

Take away the units of measurement and you end up with:

$$x + y = 15$$

Now we can figure out how much "x" is.

All I have to do is solve for x in terms of y. Do you remember that from volume III? Let me remind you. In the problem below, I want to get x by itself. I will move that blue "y" over to the other side to accomplish that.

$$x + y = 15$$

$$x = 15 - y$$

There we go! That's it! I just solved for x in terms of y. All that means is you can give me any number in the world for "y" and then I can tell you know how much "x" is. Right now, I don't know how much x or y is, but if you told me that y = 10, then I would know that x is equal to 5, according to the equation above.

Now that we know all the amounts, let's add them to our picture. Remember, "x" is equal to "15 - y," so I replaced our "x" with "15 - y."

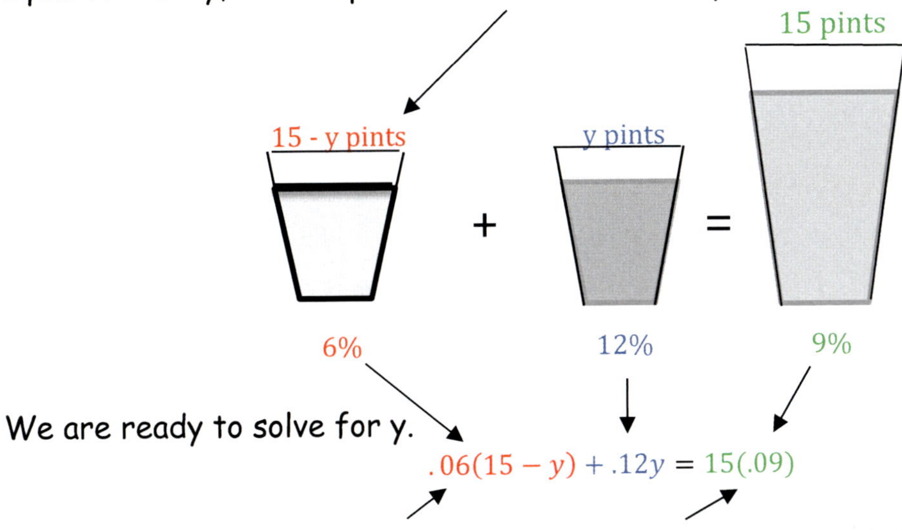

We are ready to solve for y.

$$.06(15 - y) + .12y = 15(.09)$$

BE SURE TO WRITE 6% AND 9% WITH A ZERO IN FRONT. 60% = .6, but 6% = .06

$$.06(15 - y) + .12y = 15(.09)$$

$$.9 - .06y + .12y = 1.35$$

$$.9 + .06y = 1.35$$

$$.06y = 1.35 - .9$$

$$.06y = .45 \longrightarrow y = 7.5$$

If y = 7.5, then what does x equal? Well, x = 15 - y, so let's do the math.

$$x = 15 - y$$

$$x = 15 - 7.5$$

$$x = 7.5$$

See, it's not so bad. You already had all the skills you needed to solve that word problem. Let's try another one. This time, you will know how much is in the first container, but you won't know how much is in the other two containers. Here is our problem:

A plumber has 10 cups of a solution that is 75% acid, but that is too strong. It might harm the pipes. He needs the level of acid to be at 50%. He has another container of a solution that is 10% acid. How much of this solution should he add to the 10 cups to get a solution that is 50% acid?

Do you know where to start? Let's start with the first container that holds 10 cups of a 75% acid solution.

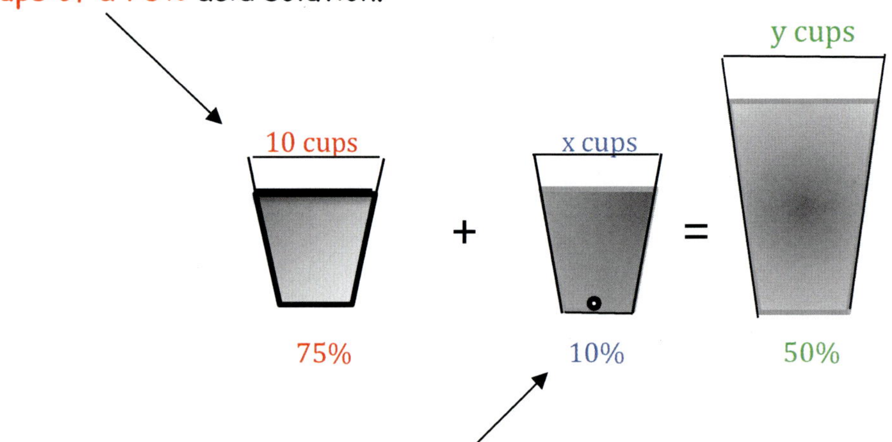

The solution in the second container is 10% acid, but we don't know how much of it we are going to add to the final container. That is our "x." The final container will hold a solution that is 50% acid, but how many cups of that solution will there be? We don't know that yet because we don't know how much "x" is. The final amount will be called "y."

This time, it is easy to see that how much "y" is equal to. Just look at the last picture. Can you figure out how much y is? It is equal to the two containers together, so y = 10 + x.

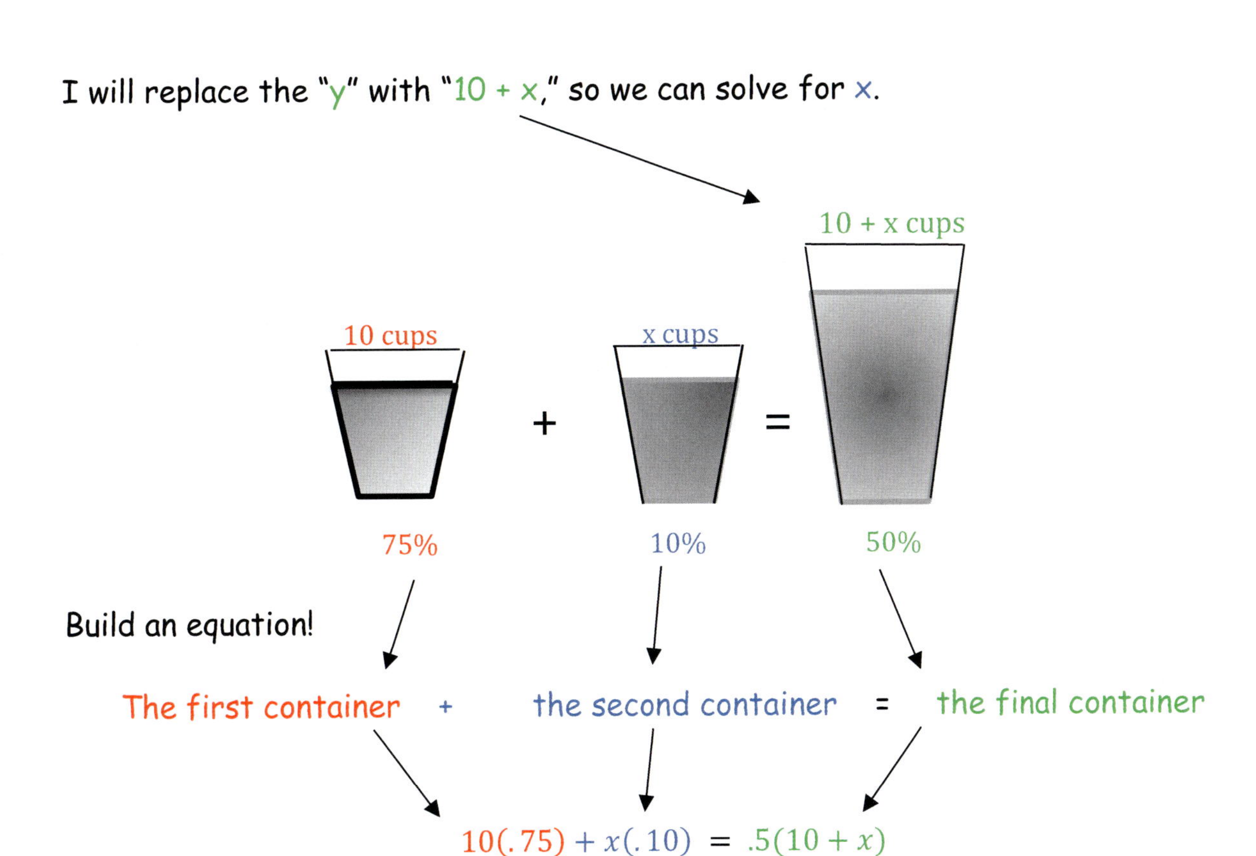

$$7.5 + .1x = 5 + .5x$$

Let's stop here for a moment. When you see this equation, you may be confused about which terms to move to which side. Keep two things in mind. You want all the "x" terms on one side of the equal sign and all the numbers on the other side. That clears up one issue, but now look at the two terms in green in the last equation above. How do you know if you should move the ".5x" to the left side or the "5" to the left side? That's a good question. I'm glad you asked.

If I were to move the ".5x" over to the left and the "7.5" over to the right side of the equal sign, then I would end up with two negative answers, like this:

$$7.5 + .1x = 5 + .5x$$
$$.1x - .5x = 5 - 7.5$$

$$-.4x = -2.5$$

I mean, it's OK, if you do, but if it were up to me, I'd move the "5" over to the left and the .1x to the right. That way my two answers are both positive numbers, like this:

$$7.5 - 5 = .5x - .1x$$

$$2.5 = .4x$$

This brings me to another point. When you get to a problem like this:

$$.5x - .1x$$

Instead of grabbing a calculator or writing it out on paper, look at those numbers as if they were money. The number ".5" is the same thing as fifty cents and the number ".1" is equal to ten cents. When I see the equation above, I see "50 cents minus 10 cents." That makes it easy for me to see that the answer is 40 cents, or .40, or .4x to be exact.

And one more thing. In the two examples above, one solution had negative numbers while the other one turned out positive, like this:

$$2.5 = .4x \qquad \text{and} \qquad -.4x = -2.5$$

They are both correct because if you solve for x in either case, the answer will be x = 6.25. But that is not my point. My warning to you is if you end up with one side positive and the other side is negative....you've probably done something wrong. Go back and figure out why your negative and positive signs are messed up.

Now, we can get back to our problem. Let me refresh your memory. Here is the problem and the math we've done, so far.

A plumber has 10 cups of a solution that is 75% acid, but that is too strong. It might harm the pipes. He needs the level of acid to be at 50%. He has another container of a solution with 10% acid. How much of this solution should he add to the 10 cups to get a solution that is 50% acid?

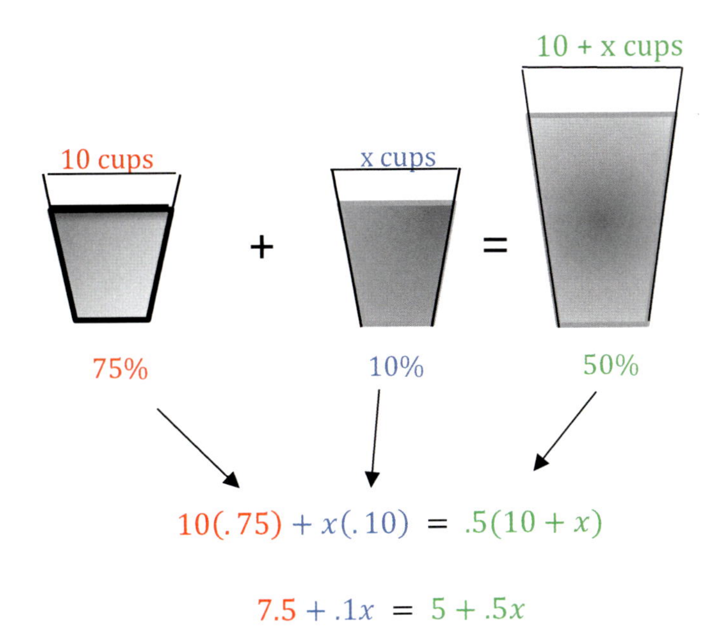

$$10(.75) + x(.10) = .5(10 + x)$$

$$7.5 + .1x = 5 + .5x$$

This is where we left off. I'll solve it my way, so we get positive numbers.

$$7.5 - 5 = .5x - .1x$$

$$2.5 = .4x$$

$$x = \frac{2.5}{.4}$$

$$x = 6.25\ cups$$

The plumber should add 6.25 cups of the 10% solution to the 10 cups to get a solution that is 50% acid.

But how do I know if I got the right answer? I can't really visualize 10 cups of this and 6 1/4 cups of that in my mind to see if it makes sense. The way to double check my work is to fill in x with my answer and then see if both sides of the equation equal each other.

$$10(.75) + x(.10) = .5(10 + x)$$
$$10(.75) + 6.25(.10) = .5(10 + 6.25)$$
$$7.5 + .625 = 5 + 3.125$$
$$8.125 = 8.125$$

Perfect, both sides are equal, so the answer is correct. Let's try another one together.

You are conducting a scientific experiment that requires 8 ounces of water with a pH level of 15%. Your first sample of water has a pH level of 4%. A second sample comes in at 20%. You decide to mix the two samples together to create the required solution. How many of each sample should you use?

Do you know where to start? Well, I know that, at the end of the day, we want 8 ounces of water with a pH level of 15%. That is our final container, so I'll just write the end of the equation now.

$$\_\_\_\_\_\_ + \_\_\_\_\_\_ = 8(.15)$$

What else do we know? We know that we have two other containers of water. One has a pH level of 4%, the other is at 20%. What we don't know is how much of each of those solutions we are going to use. Those amounts will be x and y.

$$.04x + .2y = 8(.15)$$

OK, now we have an equation with two variables (letters). We need to get rid of one of the two, so we only have to solve for one variable. Do you remember how to do that? We need to solve for x in terms of y, to find out what x equals. I'll do the math below for you. So, let's see, the total of x and y must be 8. I'll start there.

$$x + y = 8$$

Should I subtract the x or the y? I can do either one. If I subtract the x, then I'm solving for **Y** in terms of **X** and I would replace the y. If I subtract the y, then I'm solving for **X** in terms of **Y** and I would replace the x.

$$x + y = 8 \qquad or \qquad x + y = 8$$

$$y = 8 - x \qquad\qquad x = 8 - y$$

I'll go with $y = 8 - x$ and replace the y with $8 - x$.

Here is our equation now:

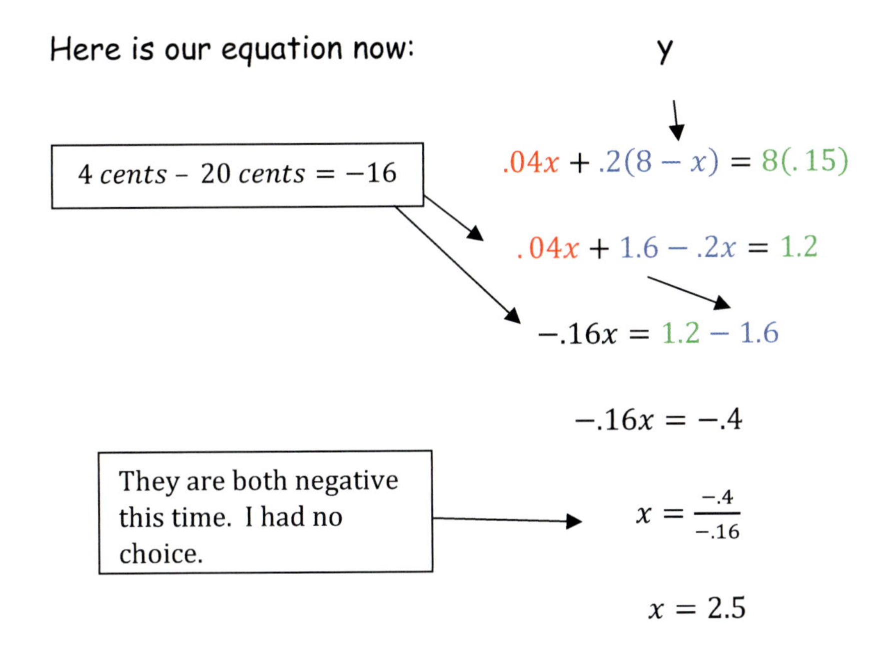

OK, so if x = 2.5, then y must equal 5.5. The final answer is 2 ½ cups of water with 4% pH level and 5 ½ cups of the 20% pH water.

In the last lesson, one of the examples had pure water as an ingredient. We had to use 0% as our level of alcohol because there isn't any alcohol in water. This time, one of our containers will hold pure fertilizer. What do you suppose the percentage of fertilizer will be in that container? That's right, it is 100% fertilizer. Here is the word problem:

A landscaper is hired to spray fertilizer on the grass at a golf course. He needs to make one gallon of a solution that is 7% fertilizer in water. How much pure water and pure fertilizer should he mix together to create the proper solution?

Do you know where to start? What is the landscaper trying to make? He needs one gallon of 7% fertilizer. That is the end of our equation. I'll write down that part first.

$$\underline{\qquad\quad} + \underline{\qquad\quad} = 1(.07)$$

So, what goes here and here? Our solution is made of pure water and pure fertilizer, but we don't know how much of each, that is our x and y.

I could draw a picture, but I don't think you need it. Just picture it in your mind. The container of pure water is 0% times x. The container full of liquid fertilizer is 100% times y. Here is the equation. Try writing the equation yourself, on a separate piece of paper, before you read my equation.

$$0(x) + 1.00(y) = 1(.07)$$

Did you write the same thing that I did? It's OK if your numbers are switched around. For example, if you wrote $x(0) + y(1) = .07 \times 1$, it is still correct. Do you remember the Communitive Property of Multiplication? That rules say that you can switch the numbers around in a multiplication problem and the answer will remain the same. That's why it is OK if your numbers are switched compared to mine.

Before we continue, I want to point out a couple things. Did you write 100% as 1? That one fools me every now and then. I want to write 100, but that is incorrect; 100% is equal to 1.00 or 1.

The other thing that concerns me is that x + y is going to equal 1, right? That means x and y are both less than one, so I will get a decimal number. And that's fine, there is nothing wrong with a decimal number, but I want my answer to be in ounces. I'm going to change the "one gallon" to "128 ounces."

$$0(x) + 1.00(y) = 128(.07)$$

Can you figure out what to put here in place of the "y?" Well, now x + y = 128, so y must equal 128 - x.

I'll add that to our equation.

$$0(x) + 1.00(128 - x) = 128(.07)$$

$$0 + 128 - 1x = 8.96$$

$$128 - 1x = 8.96$$

$$-1x = 8.96 - 128$$

$$-x = -119.04 \qquad so, x = 119.04$$

If $x = 119.04$, then $y = 8.96$. The landscaper should use 119.04 ounces of water and 8.96 ounces of fertilizer. Try this problem again on your own, but use one gallon this time, instead of 128 ounces. You will end up with a decimal number. Multiply that number by 128 and see if we got the same answer.

Did you get the same answer? Great! Complete the next worksheet. At the end of each problem, be sure to double check your work. There are a lot of calculations in each problem, so it is easy to make an error. Your answer may look correct, but you can't be sure until you replace x and y with your answers. If both sides of the equation are equal, then your answer is correct.

WORKSHEET 9

1. In this part of the country, the water in a car's radiator should be 20% anti-freeze, so it doesn't freeze during the winter months. The radiator holds 5 gallons of water and anti-freeze solution. How much pure water and pure anti-freeze should be used to obtain the proper solution in the car?

2. A nurse has two containers of Isopropyl Rubbing Alcohol. One container has 70% alcohol and the other one has 91%. She is instructed to use just enough of each solution to obtain 50 milliliters of 80% Isopropyl Rubbing Alcohol. How much of each container should she use? Round your answers to the nearest one-hundredth.

Worksheet 9 page 2 of 3

3. A nurse is preparing an IV bag of water and sugar. She has a large bottle that is 15% sugar, but the IV bag needs to be one liter of 9% sugar water. How much of the bottle should she add to what amount of water?

4. A one-gallon tub of Epsom Salt and water is 6% Epsom salt. How much pure water should be added to make a solution that is 4% Epsom salt?

Worksheet 9 Page 3 of 3

5. A solution of acid and water is 16% acid. How many cups of this solution should be mixed with pure water to make 21 cups of a solution that is 10% acid? Do not round your answer.

6. A masonry needs to clean a concrete wall. A solution of bleach and water works very well. He will need 5 gallons of a bleach water solution that is 20% bleach. He has a container of 12% bleach water. How much of that solution should he mix with pure bleach to get the proper percentage?

7. Make up your own solution problem with two variables. Solve it and then double check your work by replacing x and y with your answers. If both sides of the equal sign are equal, then your answer is correct.

LESSON 10: MIXTURE PROBLEMS

In this lesson, you will learn about one more type of problem that is sure to be found on just about any algebra math test. It is called a *Mixture Problem*. In these types of problems, there are usually two different items, with two different values, being combined to create a new *mixture* of items. For example, you might mix two different types of flowers together to create a bouquet of flowers.

In prealgebra you learned how to solve for x in terms of y, so you could plot points on a graph. (If this is the first time you've heard this, then you need to read Volume III and V of the *Learn Math Fast System*.)

In algebra, you will solve for x in terms of y to help you solve story problems. Let me explain with something that's not so "mathy." Look at the sentence below. The variable, x, will stand for a particular color.

Anita is wearing (x) shoes.

Let's say, x = blue. Now rewrite the sentence, but put the word "blue" in place of the variable.

Anita is wearing (blue) shoes.

Simple, right? Now I will add another variable to the original sentence.

(y) is wearing (x) shoes.

y = Melanie x = red

Rewrite the sentence, but substitute the variables with their assigned values. The value assigned to "y" is Melanie. The value assigned to "x" is red.

(Melanie) is wearing (red) shoes.

Still easy, isn't it? OK, then I'll make it more difficult.

Rewrite the original sentence using these values for x and y.

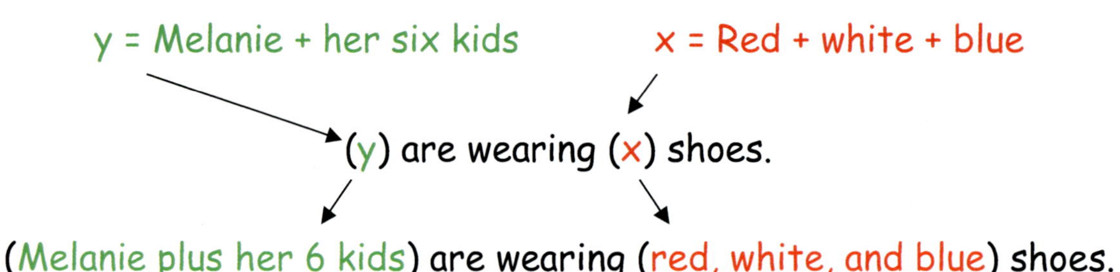

Of course, you can drop the parentheses and it will mean the same thing. Now let's take a look at the exact same situation, but this time I will use algebra instead of words.

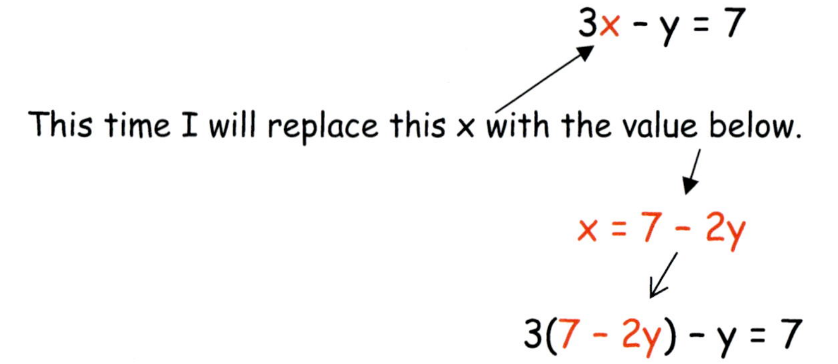

Do you see how that is the same thing as replacing the variable "x" with the word "blue?" You may be asking, "Why would I ever want to do such thing?" The reason is so you can solve for "y."

A typical mixture problem will have two variables, usually x and y. To solve these two variable problems, you will need to solve for x in terms of y, so you can then turn around and solve for y in terms of x!

I'll give you an example of a mixture problem and then we will solve it together.

Two kinds of candy sell for 45 cents and 60 cents/pound. How many pounds of each should be used to make 45 pounds of a mixture to sell for 50 cents/pound?

First of all, what does this question mean? There are two different candies. Let's say one is chocolate drops and the other is peanut butter drops. The factory would like to put together a mixture of chocolate and peanut butter drops and sell them together for 50 cents per pound.

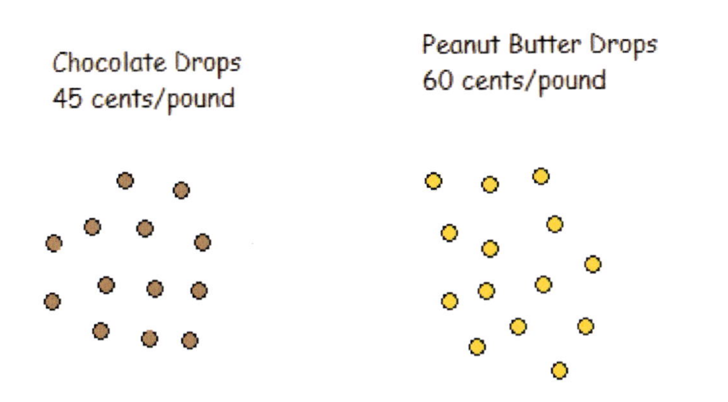

45 pound bag of Peanut Butter and Chocolate Mix

50 cents/pound

They don't want to lose money by adding too many expensive peanut butter drops to the mix, and they don't want too many chocolate ones either. They need just the right amount of each candy to make money and have a good mixture. So, they ask the math whiz (that's you) to figure out how many pounds of each candy should be poured into the big 45 pound bag that will sell for 50 cents per pound.

A mixture problem is also called a *weighted* problem because each variable has a different *weight*. Let me explain. I will call each pound of chocolate "x." Each one of those pounds has a *weight* of 45 cents. I will call each pound of peanut butter, "y." Each of those pounds has a *weight* of 60 cents. And we are trying to create a 45-pound bag *weighted* at 50 cents per pound. Here is how that looks in an equation.

$$.45x + .6y = 45(.5)$$

Let me explain further by using colors and words. We are looking for some amount of chocolate drops (x) and some amount of peanut butter drops (y) that will equal 45 pounds.

C + PB = 45 lbs.

$(x + y = 45)$

But there is more to it than that because we have to consider the price per pound too. In other words, each term is *weighted*, so we must consider that too. The chocolate drops are at 45 cents per pound, the peanut butter drops are 60 cents per pound, and we want our 45-pound bag to be at 50 cents per pound.

No problem! We will multiply each term by their price per pound.

$$C(.45) + PB(.60) = 45(.50)$$

Now each term is properly weighted. I will replace the "C" and "PB" with an "x" and "y" respectively.

$$.45x + .6y = 45(.5)$$

This is a two-variable equation. To solve it, we will need to replace one of the two variables with another value, so we end up with only one variable in our equation.

Look back at the last page. We determined that together the chocolate and peanut butter drops will equal 45 pounds, right?

$$x + y = 45$$

Alright, then let's solve for y in terms of x, so we have a value for y.

$$y = 45 - x$$

Does that equation make sense to you? Think about what each term stands for.

$$y = 45 - x$$

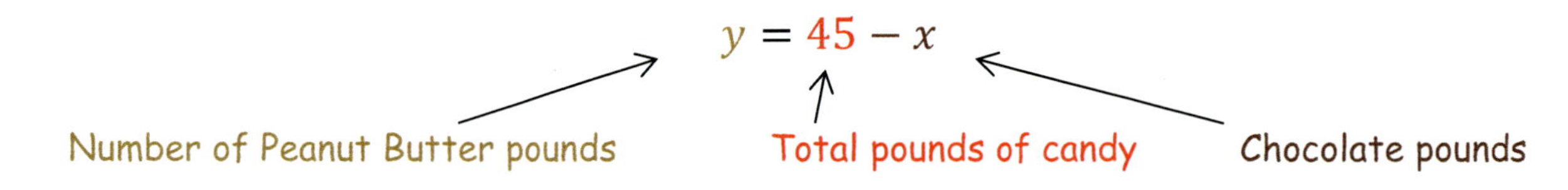

Now we can replace the "y" in our equation with (45 - x). This will turn our two-variable equation into a one-variable equation.

$$.45x + .60(45 - x) = 45(.50)$$

Now do you understand this equation? It is saying:

(45 cents/lb times "x" pounds) + (60 cents/lb times 45 lbs. minus chocolate lbs.)

$$.45x + .60(45 - x) = 45(.50)$$

Equals (45 pounds times 50 cents/lb.)

That was the hard part. Now we get to do the fun part - solve for x!

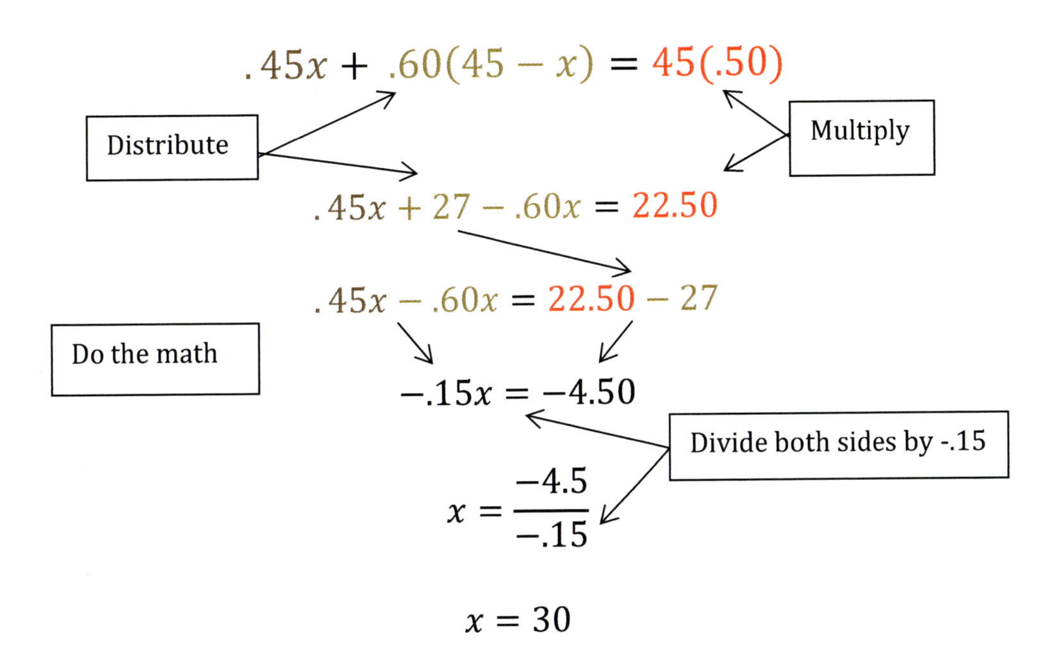

OK, so if $x = 30$, then we need 30 pounds of 45 cent candy, right? And how many pounds of the 60-cent candy?

$$.45x + .60(45 - x) = 45(.50)$$

That's right, 45 - x. And if x = 30, then we need 15 pounds of the 60-cent candy.

If you are confused, read through that last example again, very slowly. If you were able to follow along, let's try another one! Here is our next example.

A florist sells roses for $1.50 each and lilies for $1 each. How many of each should be included in an arrangement of two dozen flowers to sell for $34 if the price of greenery in the arrangement is $5?

This problem is very similar to that last problem, except with a little twist. This one has an added term, +$5, the price of the greenery. That's no problem; we can easily toss that in to the mix. Here is what we know:

Roses = $1.50 each
Lilies = $1.00 each
Greenery + $5 each arrangement

We want a mixture of 24 flowers to equal $34.

Let's bring back our chocolate and peanut butter equation for a moment. And then we can model our flower equation the same way.

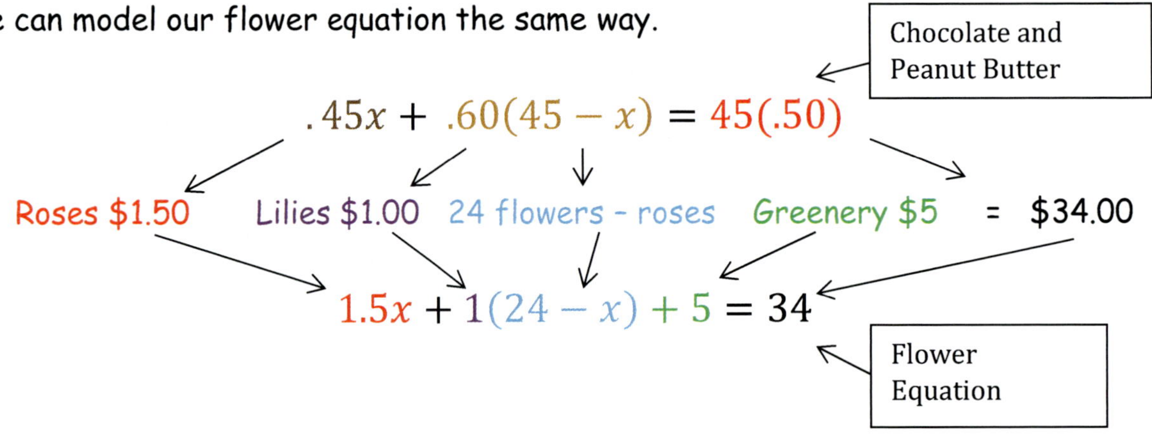

Study everything written above until you fully understand the flower equation. If you don't understand it yet, read the candy example again. If are following along just fine, let's solve the equation!

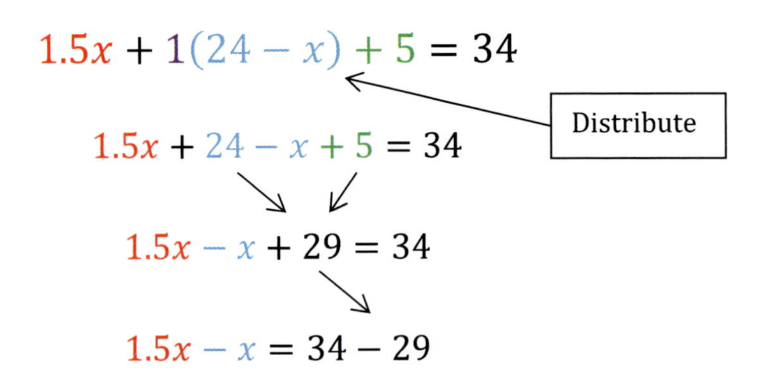

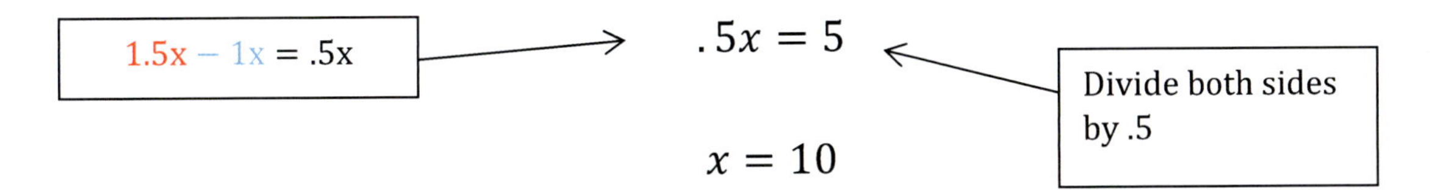

OK, so x = 10. What does that mean? I'll bring back our equation and we will READ it together.

$$1.5x + 1(24 - x) + 5 = 34$$

The term written in red, means $1.50 per rose times "x." The next term means $1.00 per lily times "24 flowers minus the number of roses." The third term is the price of the greenery in each flower arrangement. And of course, these three terms should all equal $34.00.

I said that x = 10, so let's see if I got it right. I'll bring back our equation, replace the "x" with 10, and we'll see if both sides of the equal sign are indeed EQUAL.

$$1.5x + 1(24 - x) + 5 = 34$$

$$1.5(10) + 1(24 - 10) + 5 = 34$$

This side equals that side.

$$15 + (14) + 5 = 34$$

I got it right!

YAY! I got it right!

If you fully understood those two mixture problems, complete the next worksheet. If you are completely lost, you might need to go back to Volume 3 or 5 of the *Learn Math Fast System*.

If you get stuck on this worksheet, take a peek at the answers. BUT just a peek! And then try again on your own.

WORKSHEET 10

Name ___________________________________ Date _______________

1. We have two types of Christmas cards. One set has a glitter border; they sell for 50 cents each. The other set is plain and only cost 30 cents each. How many of each should be placed in assortment boxes of 50 cards each to sell for $18 per box?

2. A dairy sells pure cream for 75 cents per cup and skim milk for 25 cents per cup. How many cups of each should be mixed to make coffee creamer that will sell for $8 per gallon?

3. A store owner is selling fresh cut fruit in $4\frac{1}{2}$ pound containers. The berries cost $4.75 per pound and the melon costs $4.20 per pound. The store owner plans to add $5 per container for the labor, materials, and profit. If he wants to sell the containers of fruit for $25 each. How many pounds of each fruit should he put in each mixture?

Name ______________________________ Date __________________

CHAPTER 2 TEST

1. A number is randomly selected from the set of numbers below.

   $$5, 11, 23, 42$$

   What is the probability of that number being greater than 25?

2. A drawer holds 12 socks. There are 6 blue socks and 6 black socks. What is the probability of selecting two socks of the same color?

Solve the inequalities.

3. $4x - 6 < 7x + 12$

4. $-6 < -2x + 4$

5. $\frac{4x+6}{2} < 4$

6. A solution of acid and water is 16% acid. How many cups of this solution should be mixed with pure water to make 32 cups of a solution that is 10% acid?

7. A solution of water and lemon flavoring will be mixed to make 8 cups of lemonade. We don't want the drink to be too sweet or too sour, so it must be 15% lemon flavoring. How much water should we use?

8. A solution of salt and water is 30% salt. How many cm^3 of this solution should be mixed with pure water to make 18 cm^3 of a solution that is 10% salt?

Chapter 2 Test page 3

9. The owner of a pizza store wants to write down the exact recipe for a pepperoni and olive pizza, so he is sure to keep the cost and weight the same for every pizza. He already knows how much a cheese pizza costs and weighs and would like to add 8 ounces of toppings for only $1.80 more per pizza. The pepperoni costs 90 cents per ounce and the olives cost 10 cents per ounce. How many ounces of each ingredient should the store owner use?

10. Debbie wants to make snack mix and then fill a 4-cup bag that cost $5.50 per bag. The cereal costs $0.50 per cup and the nuts cost $3.00 per cup. The bag, butter, and seasoning costs 1.00 per 4 cup bag. How much cereal and nuts should be added to each bag to keep the cost at $5.50?

CHAPTER 3: TWO VARIABLE EQUATIONS

LESSON 11: SYSTEMS OF EQUATIONS

This math will look really complicated to someone who doesn't understand algebra, but don't tell anybody that it's actually really easy because it makes us look smart.

First of all, do you remember learning about the slope of a line? Do you remember what a linear equation means? If these words are new to you, read Volume III of the *Learn Math Fast System*.

Below is a picture of a graph with a line drawn on it.

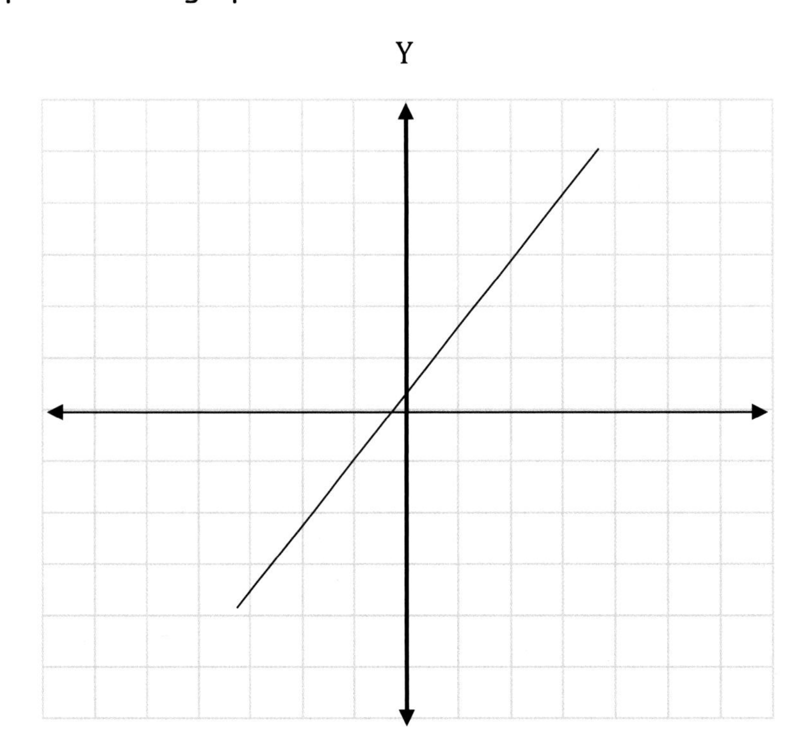

In volume III you learned how to find the *y-intercept*. That is the point at which our line crosses, or intersects, the y-axis. We used this formula to find the y-intercept:

$$y = mx + b$$

Next, I will draw another line on the same graph. When you put two lines on one graph, one of three things will happen. The two lines will be parallel to each other and never touch, they will be identical and overlap each other, or they will, at some point, intersect each other. Look below.

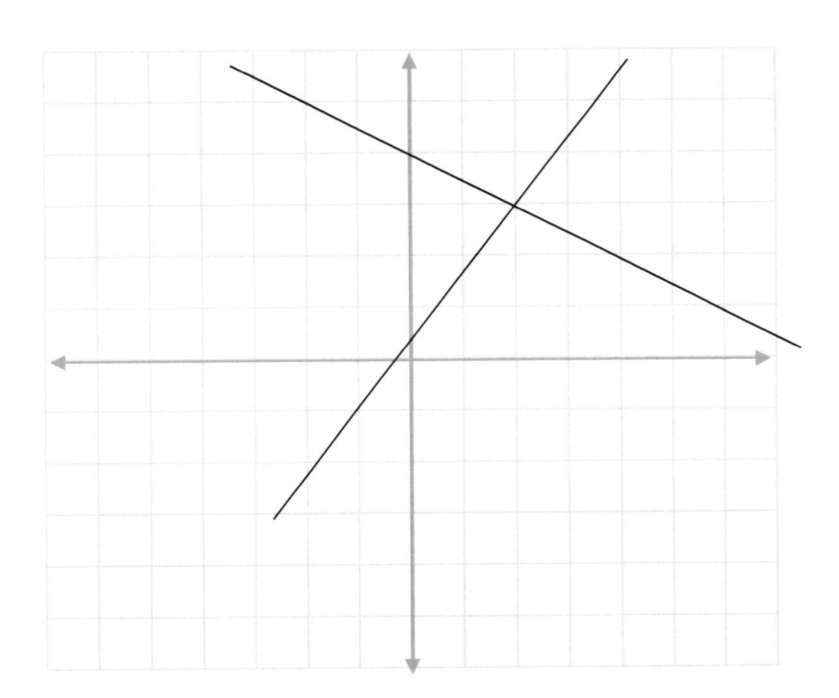

These two lines are not parallel, and they are not identical. These two lines intersect each other at one and only one point. To find the coordinate points that these two lines share is to find the *simultaneous solution*. The word *simultaneous* means "at the same time" and the word *solution* means "the answer." So, a *simultaneous solution* is the answer that will solve two different linear equations *at the same time*. Of course, you can just look at the graph above and figure out that the point these two lines share is (2, 3), but there is more to it than that.

The next level of story problems we are going to solve will require two variables in our equations. To solve this type of problem, you will need to learn how to solve for the simultaneous solution in a system of equations.

Remember, a linear equation is an equation that draws a line. The two lines above represent two different linear equations. So, this time, instead of drawing a picture for you, I'm going to give you two linear equations and we will have to mathematically figure out at which point they intercept each other. Sounds fun, doesn't it? Let's get started. Here are two linear equations:

$$x + 2y = 7$$
$$3x - y = 7$$

If you drew these two lines on a graph, they would intersect each other at one and only one point. We are going to figure out that point by doing some fancy math. In "mathy" terms, we are going to find the *simultaneous solution of these two linear equations*. To do that, we add or subtract (combine) the two equations together, but we do it in a sneaky way. First, we change one of the two equations, so when we add them together, one of the two variables will go away. That will make it easier to solve. It's OK if you don't understand this yet; I'm going to show you all the steps.

Before we can add these two equations together, we need to get rid of one of the two variables. It doesn't matter which one, so let's take a look at the two equations and figure out which one would be easiest to eliminate. Keep your eye on this term, $+2y$.

$$x + 2y = 7$$
$$\underline{+3x - y = 7}$$

I have decided to get rid of that term, that way I only have to deal with one variable, x. To get rid of $+2y$ I will need to "minus 2y" during our addition work. The lower equation has a like term $-1y$. But we want it to be $-2y$, so it cancels out the $+2y$ in the equation above. That is no problem, I can change that.

I can multiply both sides of that lower equation by 2. That will automatically make that $-y$ turn into $-2y$. Look at the math below.

$$x + 2y = 7$$
$$3x - y = 7$$
$$\downarrow \qquad \searrow$$
$$2(3x - y) = 2(7)$$
$$\downarrow \quad \downarrow \quad \downarrow$$
$$6x - 2y = 14$$

There we go, now we have a $-2y$ in our equation. Now when we add these two linear equations together, our little friend "2y" goes away. I have added the two equations together, below.

$$\begin{array}{r} x + 2y = 7 \\ \underline{+6x - 2y = 14} \\ \mathbf{7x \qquad = 21} \end{array}$$

$\mathbf{x + 6x = 7x}$ $\qquad$ $\mathbf{+2y - 2y = 0}$ $\qquad$ $\mathbf{7 + 14 = 21}$

By adding the two equations together, not only did we get rid of that "2y" we also shrunk our equation down to $7x = 21$. Let's complete that equation by solving for x.

$$7x = 21$$

Divide both sides by 7

$$x = 3$$

OK, now we have a value for "x." Remember, we are trying to find a set of coordinate points (x, y) that both of our linear equations share. The answer above is the first of the two coordinates. Now we can go back to the original equations and figure out a value for "y".

The original equations I gave you are below. We will work with the one on top first.

$$x + 2y = 7$$

$$3x - y = 7$$

Since we know that x = 3, plug that into one of the equations above and solve for y. I'm using the top one.

$$3 + 2y = 7$$

$$2y = 7 - 3$$

$$2y = 4$$

$$y = 2$$

Now we have a value for x and y (3, 2) that will work for both linear equations. That means, if we were to draw the two lines that were created by those linear equations, they would intersect each other at (3, 2) on a graph.

Does that make sense to you? I'll review what I just did. We were given two linear equations. We were told that these two lines will intersect each other when drawn on a graph. To find this point, we had to add the two equations together. Before we added them together, we changed one of the equations to make sure that one of the variables would go away, making it easy to solve for x.

Once we had a value for x, we plugged that into one of the equations, to solve for y. This math gave us a simultaneous solution; the one and only point where the two lines cross each other's path.

Let's do that again with the same two equations. Only this time we will select a different term to make disappear.

Our two original equations are below.

$$x + 2y = 7$$
$$3x - y = 7$$

This time, we will start by getting rid of the $+3x$. To get rid of "positive 3x" we need to have a "negative 3x" in the upper equation before we can add them together. The way to get a $-3x$ into the first equation is to multiply both sides of the equal sign by "-3."

$$-3(x + 2y) = -3(7)$$
$$-3x - 6y = -21$$

Now our equation is prepared and ready to be added, or combined, to the other one.

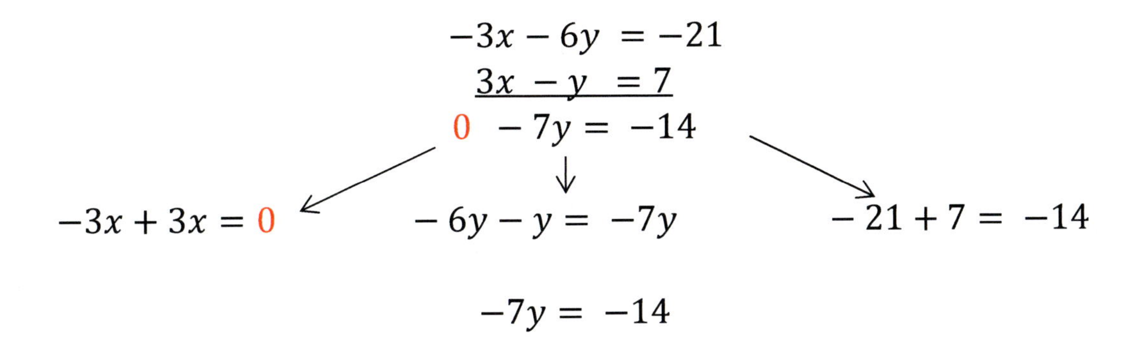

$$-7y = -14$$

Divide both sides by -7 to get y by itself.

$$y = 2$$

Now that we have solved for y, we can plug that number into one of the original equations and then solve for x.

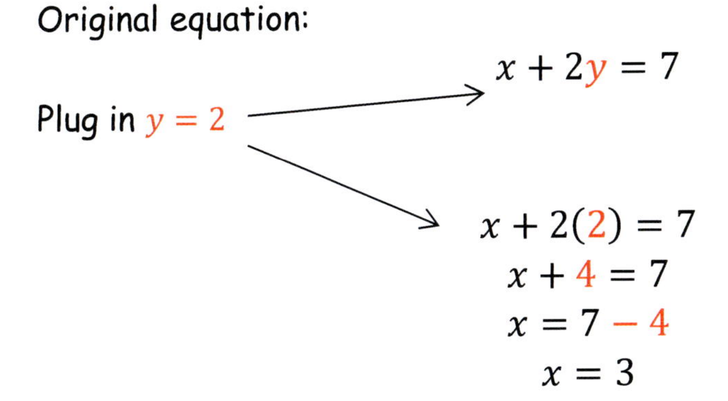

$$x + 2(2) = 7$$
$$x + 4 = 7$$
$$x = 7 - 4$$
$$x = 3$$

Look at that, we got the same answer (3, 2). You just learned how to find the *Simultaneous Solution in a System of Equations.* That is the technical way to say, you added two equations together to solve for x and y, so you could figure out where those two lines would intersect, if you drew them on a graph.

That was an easy one, so let's try a more challenging one.

Your next two linear equations are written on the next page. I want to know where on a graph these two lines intersect. You could draw a table, find some coordinates for each equation, draw a graph, and then draw the lines, to find the point at which they intersect. But I think you'll find that adding the two equations together is much faster than drawing the graph. Add the two linear equations below, to find the simultaneous solution.

$$3x + 4y = 4$$

$$2x - \frac{2}{3}y = 1$$

OK, let's see, we need to get rid of one of these terms. That's a hard choice, but I've got a plan. I'm going to get rid of the $+4y$. I'll need to turn the $-\frac{2}{3}y$ into a $-4y$ to accomplish that.

No problem, that's easy! Do you know what number to multiply by $-\frac{2}{3}y$ that will equal $-4y$?" Well maybe not right off the top of your head, but it's easy to figure out.

First, answer this question: what number times 5 will equal 15? How will you figure that out? Simple division will tell you the answer is 3. We will approach our equation the same way.

$$-4 \div -\frac{2}{3} =$$

Do you know how to divide this problem? Look at it as fractions. I'll rewrite it below.

$$-\frac{4}{1} \div -\frac{2}{3} =$$

Do you recall how to divide fractions? You flip the second fraction (get the reciprocal) and then multiply.

$$-\frac{4}{1} \times -\frac{3}{2} =$$

A negative number times a negative number has a positive answer, so you know the answer will be positive. I'll multiply straight across.

$$-\frac{4}{1} \times -\frac{3}{2} = \frac{12}{2} = 6$$

After I multiplied, I got $\frac{12}{2}$ and that reduced down to 6, positive 6.

OK, now we have a number that we can multiply by both sides, to get rid of the $+4y$, (finally). Let's start over, as if we knew "6" would be the best number to use. I will color the like terms the same color.

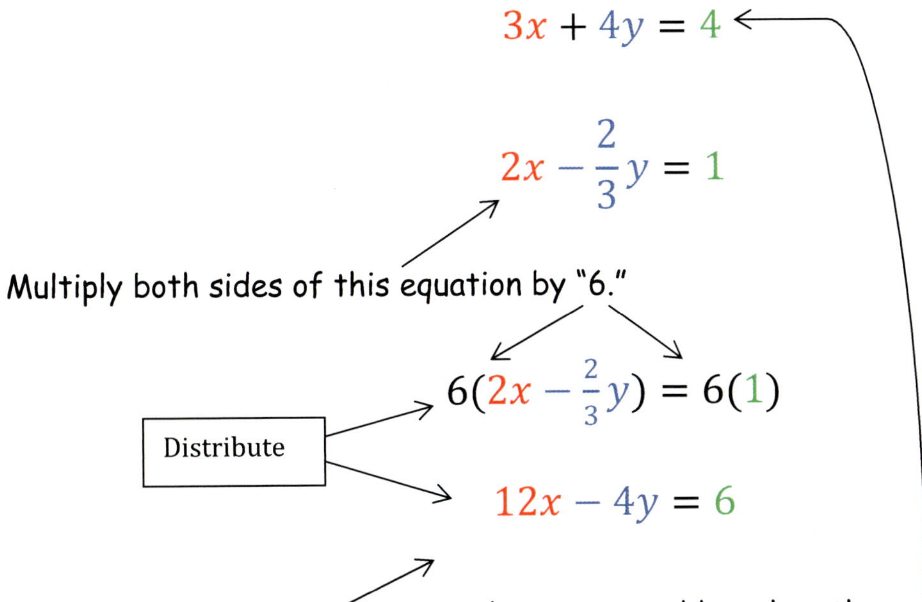

Now we finally have an equation that we can add to the other one. I've written them both below and kept the like terms the same colors. Now let's combine them.

$$\begin{array}{r} 3x + 4y = 4 \\ \underline{12x - 4y = 6} \\ 15x \quad\quad = 10 \end{array}$$

These two green numbers are both positive, so we add them together.

From here, we divide both sides by 15, to get "x" by itself.

$$\frac{15x}{15} = \frac{10}{15}$$

$$x = \frac{2}{3}$$

OK, that's the answer to x. Don't worry about it being a fraction. That just means that the x coordinate (technically called the *abscissa*) is somewhere between 0 and 1.

Now we can replace the "x" in one of our equations with $\frac{2}{3}$ to solve for "y." Here is one of our original equations.

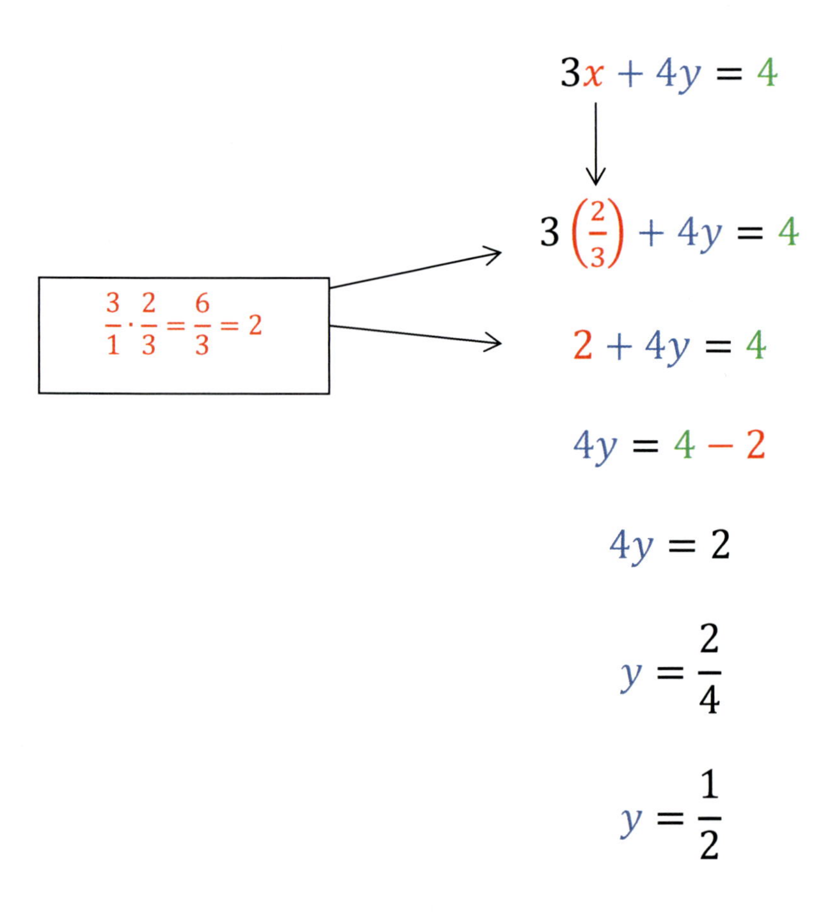

TADA! Now we have the "y" coordinate (technically called the *ordinate*) of our point. Below are the answers to both x and y. Let's fill in a set of coordinates.

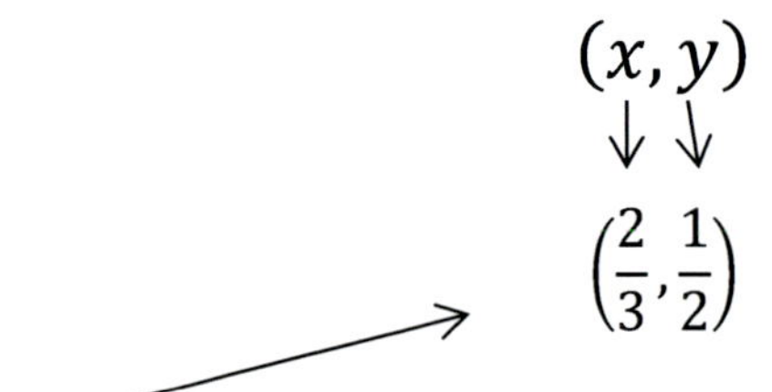

That is the point on the graph where the two lines drawn by our linear equations will intersect.

Are you wondering why anybody would ever want to know such a thing? Well there is a very good reason. Sometimes when you are building an equation to solve a story problem, you will end up with two variables in a single equation. To solve a problem with more than one variable, you will have to solve for a simultaneous solution.

But, before I have you complete a worksheet, let me show you another way to solve a system of equations. This time we will use the "*Substitution Method*" to solve for the simultaneous solution. Some people find this way to be easier. Here is our first example:

$$9x - y = 1$$

$$3x + 5y = 11$$

Above are two linear equations. With the *substitution* method, we will actually *substitute* one of the variables. I'll explain what I mean. First, focus your attention on the red linear equation from above. I selected to use that equation because "y" doesn't have a coefficient, so it's nice and easy to solve for "y."

$$9x - y = 1$$

I am going to solve for y in the equation above.

$$-y = 1 - 9x$$

I don't want to leave this as a *negative* y. So in order to change that sign, we will have to multiply –y by -1. That will make it positive. And whatever you do to one side of the equation you must do to the other side, right?

$$-1(-y) = -1(1 - 9x)$$

$$y = -1 + 9x$$

OK, now we have a value for y. We are going to substitute the "y" in the green equation above with our y value.

$$3x + 5y = 11$$

$$\downarrow$$

$$3x + 5(-1 + 9x) = 11$$

Now that we have gotten rid of the "y" variable, it is time to solve for x.

$$3x + 5(-1 + 9x) = 11$$

$$3x - 5 + 45x = 11$$

$$48x = 11 + 5$$

$$48x = 16$$

$$x = \frac{16}{48} \quad or \quad x = \frac{1}{3}$$

OK, now we have a value for "x." Let's bring back our red equation and replace the "x" with $\frac{1}{3}$.

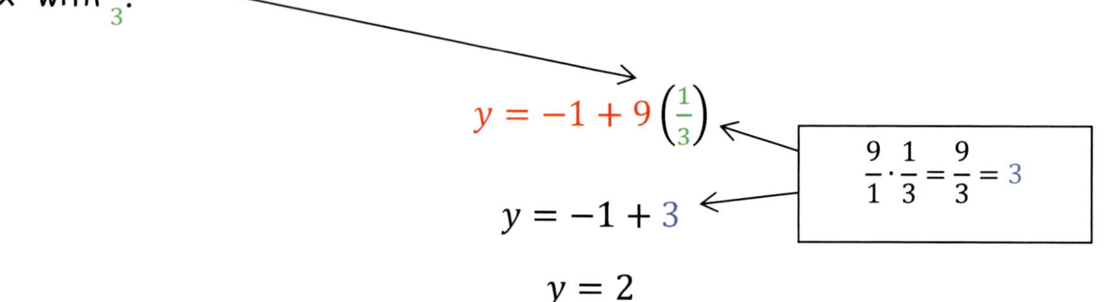

And there you have it. With the substitution method we found that $x = \frac{1}{3}$ and $y = 2$. Use the method of addition that we learned earlier to solve this problem on your own. Did you get the same answers that I did with the substitution method?

In the next lesson, you will learn how to solve story problems by using these systems of equations. But first, complete the next worksheet to practice finding the simultaneous solution of two linear equations. You can use either method; substitution or addition.

WORKSHEET 11

Name ______________________________ Date ____________

Add the linear equations together to find the simultaneous solution set of each or if you prefer, use the substitution method.

1. $x + 2y = 6$
   $2x - 3y = 5$

2. $3x + 2y = -2$

   $5x - y = 14$

3. $\frac{1}{2}x + y = \frac{1}{2}$
   $9x - 6y = 1$

4. $5x - 4y = 1$
   $10x + 6y = 9$

LESSON 12: GRAPHING LINEAR EQUATIONS

In the last lesson, you found the simultaneous solution of two linear equations. In this lesson, you will graph those linear equations to further prove that our answers were indeed the points at which our lines intersected on a graph.

I am going to bring back each one of the examples from the previous lesson and then we will graph each linear equation.

The first example we solved together had these two linear equations.

$$x + 2y = 7$$

$$3x - y = 7$$

We added these two linear equations together and decided that:

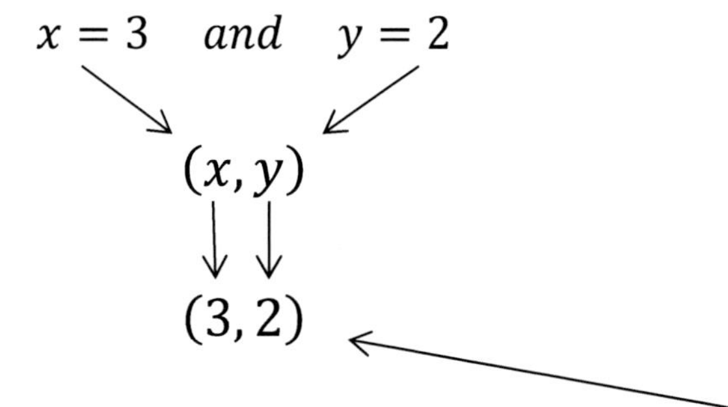

If we're correct, then the lines drawn by our equations will intersect at (3,2). To graph these lines, we will need to find several points for each equation.

Do you recall learning about slopes? Do you remember making a table, assigning some values to "x" and then solving for "y?" We are going to do the same thing here.

First I will create a table for the upper linear equation and then I will make up a few numbers for "x" to become.

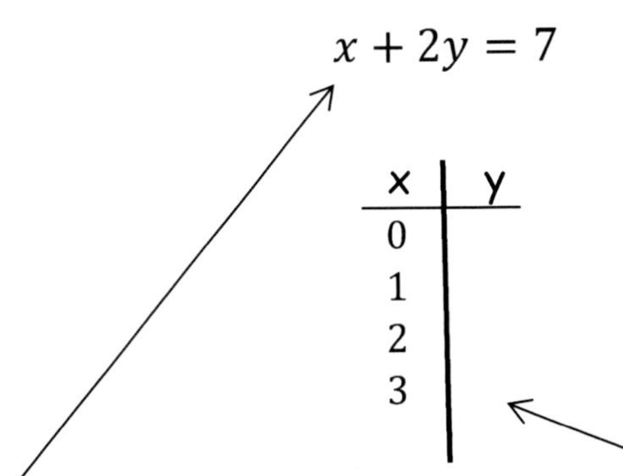

$$x + 2y = 7$$

x	y
0	
1	
2	
3	

Let's replace the "x" in our equation with each one of these numbers and then solve for "y." First will be 0. I have replaced the "x" with 0.

$$0 + 2y = 7$$
$$2y = 7$$
$$y = \frac{7}{2}$$

Now we can add this value for "y" to our table, next to the 0. Let's move on to 1.

x	y
0	$\frac{7}{2}$
1	
2	
3	

I will replace the "x" in our equation with the number 1 and then solve for y.

$$1 + 2y = 7$$
$$2y = 7 - 1$$
$$2y = 6$$
$$y = 3$$

I'll add that value for "y" to our table too.

x	y
0	$\frac{7}{2}$
1	3
2	
3	

Now it's your turn. Replace the letter "x" with the number 2. Then read my work below, to see if we got the same answer.

$$2 + 2y = 7$$
$$2y = 7 - 2$$
$$2y = 5$$
$$y = \frac{5}{2}$$

Did you get $y = \frac{5}{2}$ also? I'll add that to our table and then we will solve for "y" when $x = 3$.

x	y
0	$\frac{7}{2}$
1	3
2	$\frac{5}{2}$
3	2

$$3 + 2y = 7$$
$$2y = 7 - 3$$
$$2y = 4$$
$$y = 2$$

Now we have four sets of coordinate points that we can plot onto a graph. They are $\left(0, \frac{7}{2}\right), (1, 3), \left(2, \frac{5}{2}\right)$, and $(3, 2)$. I have plotted those points on the next page.

$x + 2y = 7$

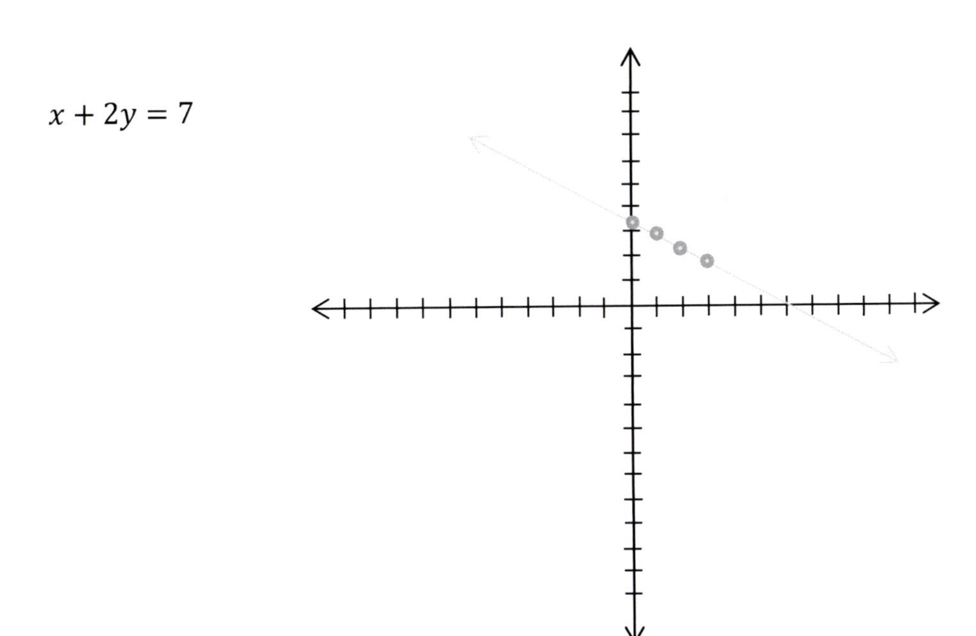

I plotted the points onto the graph above and connected the dots. Next, we will create some coordinates points for the second equation and then plot those points onto the graph too.

$$3x - y = 7$$

$x = 1$	$x = 2$	$x = 3$	$x = 4$
↙	↙	↙	↙
$3(1) - y = 7$	$3(2) - y = 7$	$3(3) - y = 7$	$3(4) - y = 7$
$3 - y = 7$	$6 - y = 7$	$9 - y = 7$	$12 - y = 7$
$-y = 7 - 3$	$-y = 7 - 6$	$-y = 7 - 9$	$-y = 7 - 12$
$-y = 4$	$-y = 1$	$-y = -2$	$-y = -5$
$y = -4$	$y = -1$	$y = 2$	$y = 5$
$(1, -4)$,	$(2, -1)$	$(3, 2)$	$(4, 5)$

I will plot the 4 coordinate points from the last page onto our graph. If all goes well, the two lines should intersect at (3, 2).

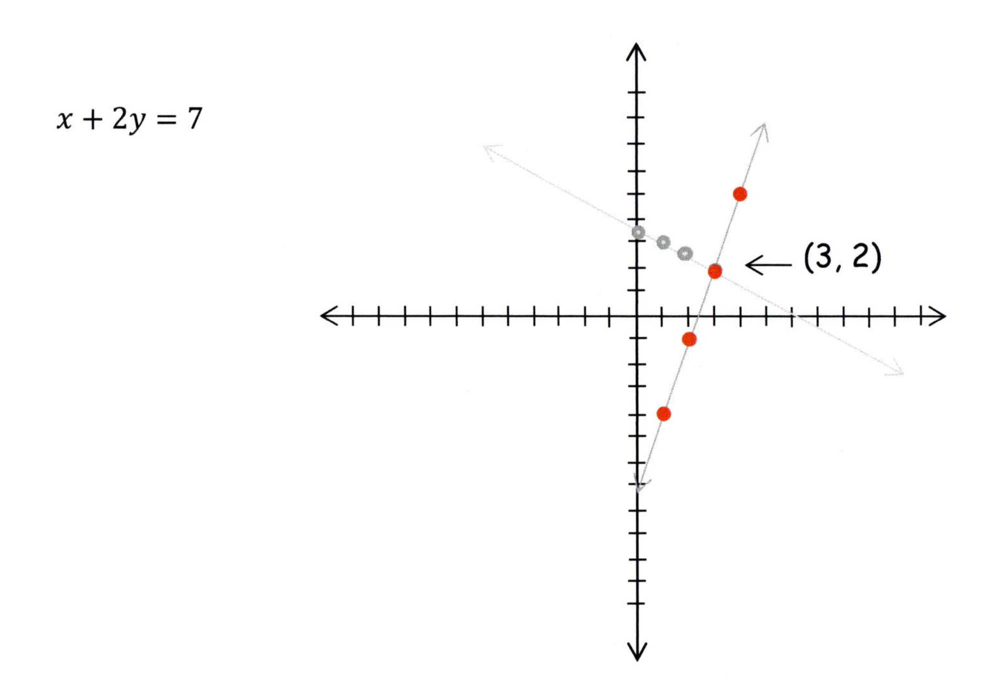

Well, would you look at that? The two lines drawn by our linear equations do indeed intersect at (3, 2). That is exactly what we came up with when we used systems of equations to solve for a simultaneous solution. In the next lesson, we will solve story problems using systems of equations. You won't need to draw any lines or graphs. The problems will be solved by creating linear equations.

Let's try this again using the second example from the last lesson. Those two linear equations are written below. We claim that $\left(\frac{2}{3}, \frac{1}{2}\right)$ is the point at which these two lines will intersect. Let's find out if we were right.

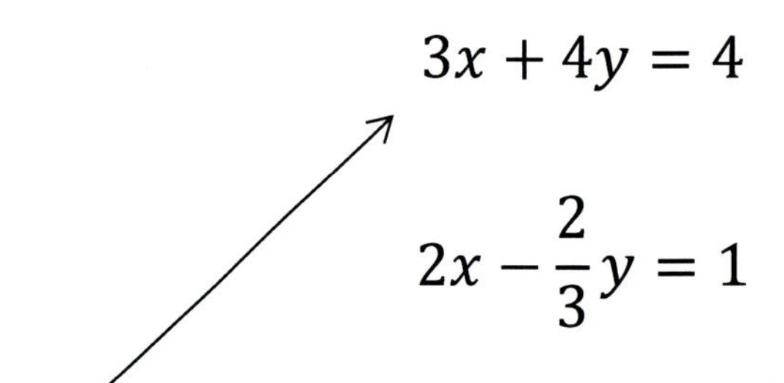

$$3x + 4y = 4$$

$$2x - \frac{2}{3}y = 1$$

I'll start with this one. I have drawn a table and gave some values for x.

$$3x + 4y = 4$$

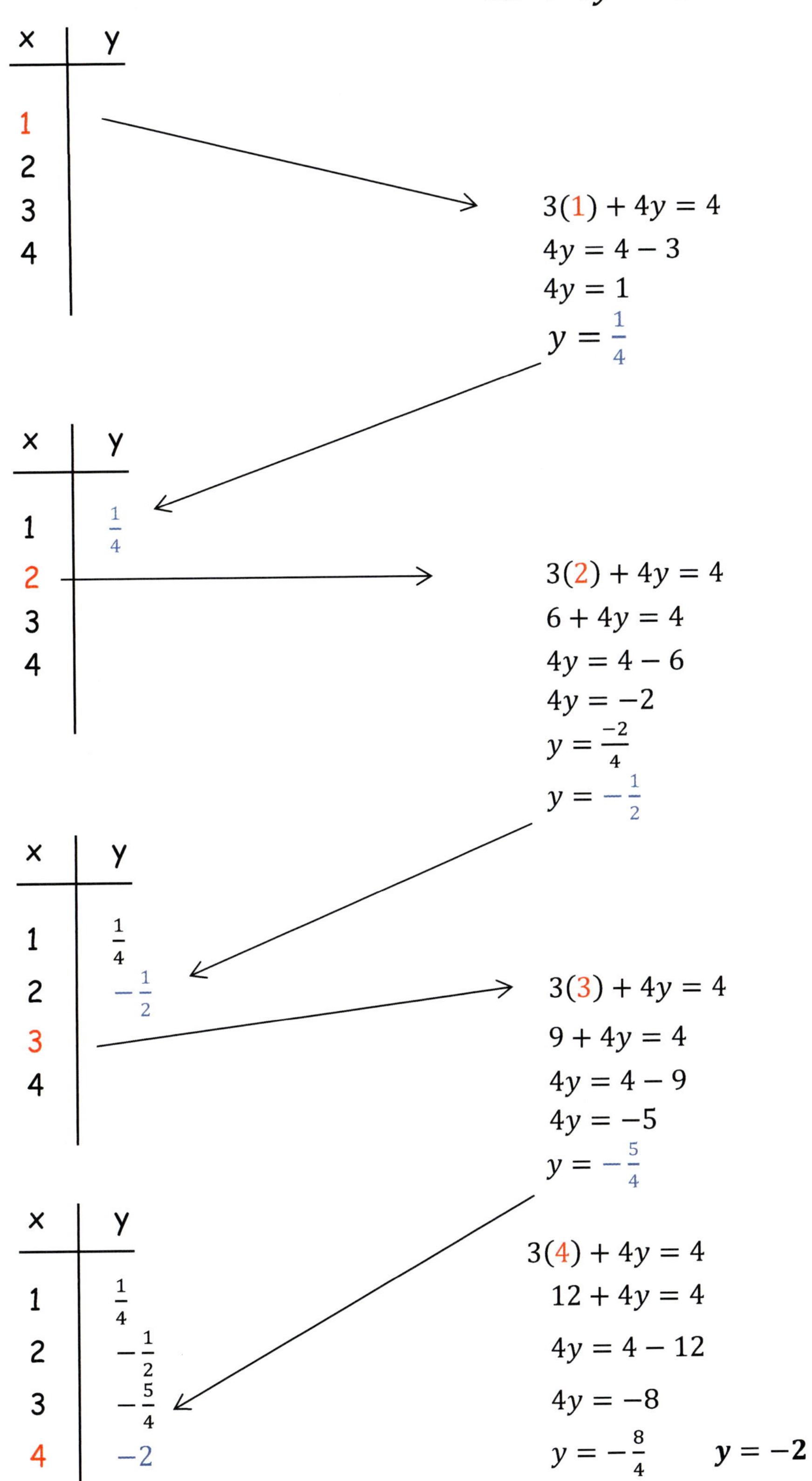

x y
1
2
3
4
3(1) + 4y = 4
4y = 4 − 3
4y = 1
y = 1/4
x y
1 1/4
2
3
4
3(2) + 4y = 4
6 + 4y = 4
4y = 4 − 6
4y = −2
y = −2/4
y = −1/2
x y
1 1/4
2 −1/2
3
4
3(3) + 4y = 4
9 + 4y = 4
4y = 4 − 9
4y = −5
y = −5/4
x y
1 1/4
2 −1/2
3 −5/4
4 −2
3(4) + 4y = 4
12 + 4y = 4
4y = 4 − 12
4y = −8
y = −8/4
y = −2

Here are the four coordinate points we created: $\left(1,\frac{1}{4}\right),\left(2,-\frac{1}{2}\right),\left(3,-\frac{5}{4}\right), and\ (4,-2)$.
Mmm…I don't see the point that we were looking for. When we found the simultaneous solution of these two linear equations in the last lesson, we got $\left(\frac{2}{3},\frac{1}{2}\right)$.
We will have to create one more set of points. This time, let's say $x=\frac{2}{3}$ and then we'll see if that will make $y=\frac{1}{2}$.

$$x=\frac{2}{3} \qquad 3x+4y=4$$

$$3\left(\frac{2}{3}\right)+4y=4$$
$$2+4y=4$$
$$4y=4-2$$
$$4y=2$$
$$y=\frac{2}{4}$$
$$y=\frac{1}{2}$$

There it is! When I set x to equal $\frac{2}{3}$ in our linear equation, y does equal $\frac{1}{2}$!
I will add those points to our list, plot them on a graph and connect the dots.

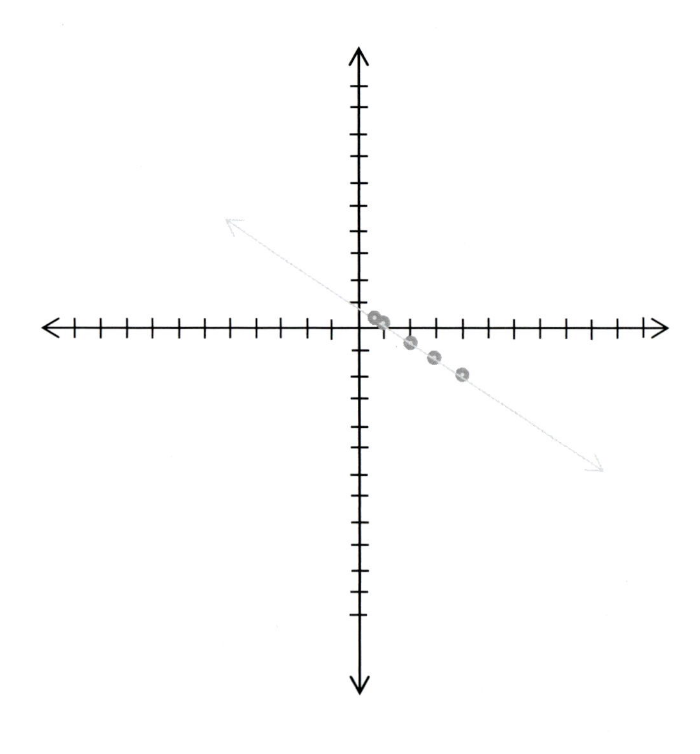

This time, you find the coordinate points for the second equation. Here it is:

$$2x - \frac{2}{3}y = 1$$

I used the numbers $\frac{2}{3}$, 1, 2, and 3, as my values for x, you should use the same. These are the points I came up with, compare them to your answers.

$$\left(\frac{2}{3}, \frac{1}{2}\right)\left(1, \frac{3}{2}\right), \left(2, \frac{9}{2}\right), \left(3, \frac{15}{2}\right)$$

Next, I will plot those points onto our graph, connect the dots, and then see where the two lines intersect.

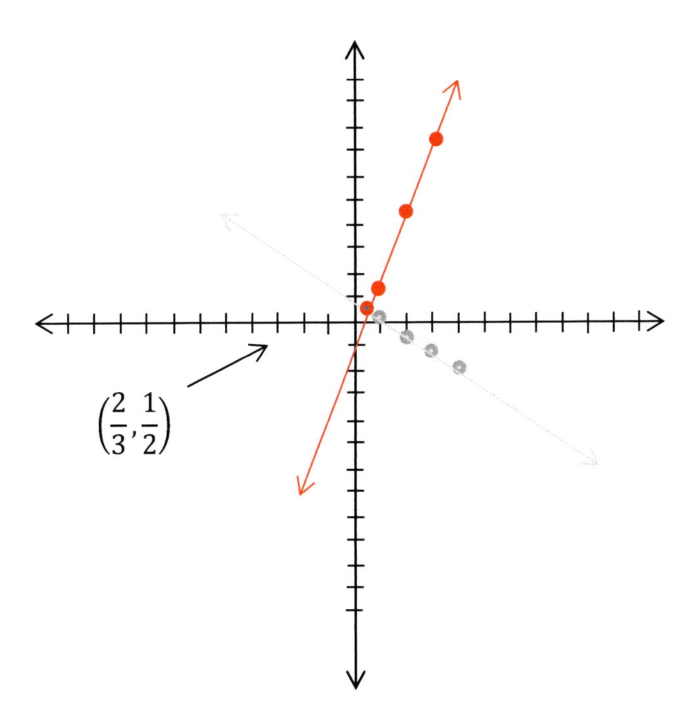

We did it! Those two lines intersected at $\left(\frac{2}{3}, \frac{1}{2}\right)$. This proves that you can find the intersecting point of two linear equations by combining them.

In the next lesson, you will solve story problems that will require you to create two linear equations. When you combine those two equations, you will get the only two possible answers that will solve that story problem. You can get the same answers by using the substitution method. Use whichever one makes the most sense to your brain.

WORKSHEET 12

Name ______________________________________ Date _________________

Below are two linear equations. Find the simultaneous solution twice, by using both the substitution method and by adding the equations together. Once you have solved the simultaneous solution, find two other points for each line using the tables below. Then, plot those two lines onto the graph to prove that you have the correct answer.

$$5x - 4y = 1$$
$$10x + 6y = 9$$

x	y
1	
$\frac{3}{5}$	
2	

x	y
1	
$\frac{3}{5}$	
2	

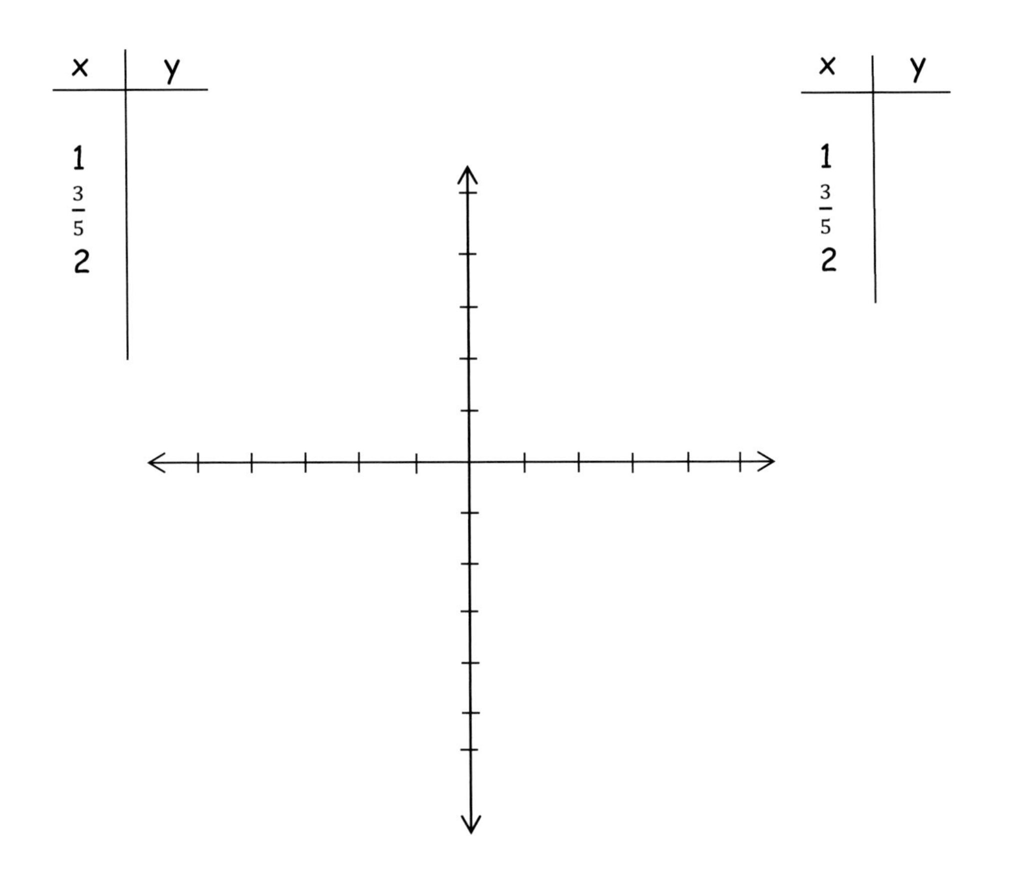

LESSON 13: SOLVING STORY PROBLEMS WITH SYSTEMS OF EQUATIONS

OK, now that you can solve systems of equations, let's solve some story problems with two variables. Below is our first example. This type of story problem always shows up on higher level math tests.

The sum of two numbers is 94 and their difference is 34. What are those two numbers?

We are looking for 2 different numbers. We will call them x and y. Can you make two linear equations with this information? The problem states that the sum of these two numbers will be 94. Do you know what a "sum" is? That is the answer to an addition problem. So that means that x + y will equal (sum up to) 94. And the word "difference" means that 34 is the answer you'll get when you put the two unknown numbers in a subtraction problem. Can you write the two linear equations on your own? Give it a try and then look at the math I've done to see if we got the same answers.

$$x + y = 94$$
$$x - y = 34$$

Did you come up with the same two linear equations that I did? It's OK if you switched around the x and the y, the answers will be the same. Now we can combine the two equations. And hey, look at that! We don't have to do any fancy work to eliminate one of the variables. It looks like the "y" variable is going to take care of itself.

$$\begin{aligned} x + y &= 94 \\ \underline{x - y} &\underline{= 34} \\ 2x \quad &= 128 \end{aligned}$$

$$x = \frac{128}{2}$$

$$x = 64$$

Now that we have a value for x, let's replace the x in one of our equations and solve for y.

$$64 + y = 94$$

$$y = 94 - 64$$

$$y = 30$$

The only two numbers that will satisfy our story problem are 30 and 64.

Now let's use the substitution method, to see if we get the same answers. I'll bring back the last two linear equations from our story problem.

$$x + y = 94$$
$$x - y = 34$$

Let's solve for y and then replace the y in one of the equations above with that value. Try it by yourself before you read my solution.

$$y = 94 - x$$

$$x - (94 - x) = 34$$
$$x - 94 + x = 34$$
$$x + x = 34 + 94$$
$$2x = 128$$
$$x = 64$$

$$y = 94 - 64$$
$$y = 30$$

Tada! I got the same answers with both methods. Let's try another one. This one will seem difficult, but I will show you how to easily solve it next.

The length of a rectangle is 5 meters longer than its width. If the perimeter of the rectangle is 54 meters, what are the dimensions of the rectangle?

Let's start by drawing a picture. We know it is a rectangular shape with a length and a width.

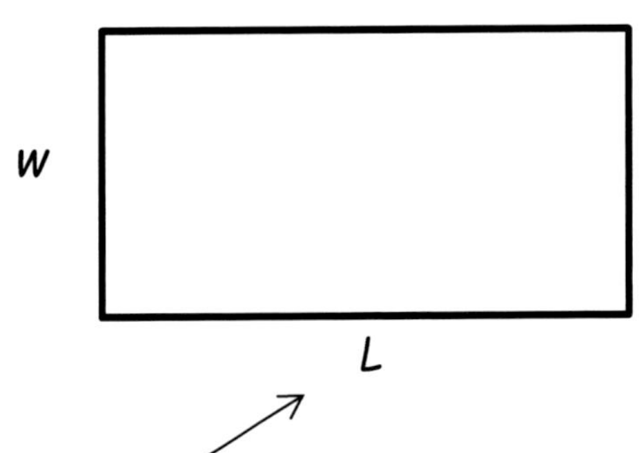

The problem says that the length is 5 meters longer than the width. I don't know what the width is, so I will call that "x." And the length is 5 meters longer.

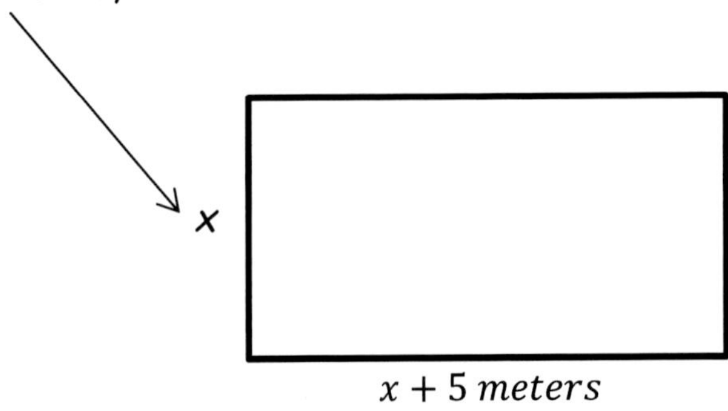

Alright, what else do we know about this rectangle? We know that the perimeter is 54 meters. And how do you solve for perimeter? The perimeter is the distance all the way around the rectangle. So, the perimeter is the width times 2, plus the length times 2, right?

$$Perimeter = 2x + 2y$$

x

y

Let's fill in this formula with the values from our story problem. The perimeter of our rectangle is 54 meters.

$$54\ meters = 2x + 2y$$

Ah ha! There is our first linear equation for this problem. We need one more linear equation, so we can add them together to find out the one and only set of answers for x and y that will solve our little rectangular problem.

The other linear equation needs to be either "y" equals something or "x" equals something. Look at the two pictures of rectangles and then try to fill in the blank below.

$$y = \_\_\_\_\_$$

Did you get it? "Y" is the length. What is our length? It is $x + 5\, meters$.

$$y = x + 5$$

There we go! Now we have two linear equations. Let's add them together, so we can find the only two numbers that will work for both x and y.

$$54\, meters = 2x + 2y$$
$$y = x + 5$$

Mmm…these aren't in the same order, so they don't line up right. Let's do a little algebra to make these two linear equations line up the same.

$$54\, meters = 2x + 2y$$

This one can just be flipped around.

$$2x + 2y = 54\, meters$$

But I'll have to minus the "x" from both sides of the next equation to get the like terms to line up with the other equation.

$$y = x + 5$$
$$-x + y = 5$$

OK, now we have two linear equations that we can try to add together.

$$2x + 2y = 54$$
$$\underline{-x + y = 5}$$

What do we do first? We need to eliminate one of the variables. Since this term is positive and this one is negative, let's get rid of the "x" terms.

$$2x + 2y = 54$$
$$\underline{-x + y = 5}$$

In order to get rid of this $-x$, we will need to multiply both sides of the equal sign by 2. I have done that below.

$$2(-x + y) = 2(5)$$
$$-2x + 2y = 10$$

There we go! Our equation is properly prepared for us to add it to the first equation.

$$2x + 2y = 54$$
$$\underline{-2x + 2y = 10}$$
$$4y = 64$$

Divide both sides by 4 to solve for y.

$$y = 16$$

Alrighty, so we've got a value for "y." Now let's get a value for x. Replace the "y" in either equation with 16, to solve for x.

$$2x + 2y = 54$$

$$2x + 2(16) = 54$$

$$2x + 32 = 54$$

$$2x = 54 - 32$$

$$2x = 22$$

$$x = 11$$

And now we have a value for x. Let's bring back our rectangle and figure out the lengths of each side.

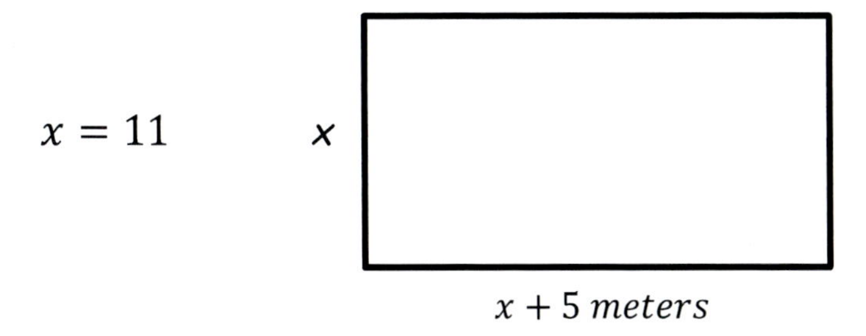

Well, that's easy! It looks like our rectangle is 11 meters wide by 16 meters long.

Just for fun, let's plot those two linear equations onto a graph. That will prove that not only are 11 and 16 the only numbers that will solve our story problem, but it will also prove that (11, 16) is the point at which those two lines will cross paths, on a graph.

We will need to find a few coordinate points for each equation, so we know where to put our dots. Since we already know that the two lines will intersect at (11, 16), let's use the numbers 9, 10, 11, 12, and 13 for our "x values." Try it by yourself on a separate piece of paper and then look at my solutions, to see if we got the same answers. Here are the two linear equations, plot them on a graph.

$2x + 2y = 54$ $\qquad$ $-x + y = 5$

$x = 9$

$2x + 2y = 54$	$-x + y = 5$
$2(9) + 2y = 54$	$-9 + y = 5$
$18 + 2y = 54$	$y = 5 + 9$
$2y = 54 - 18$	$y = 14$
$2y = 36$	
$y = 18$	

$x = 10$

$2x + 2y = 54$	$-x + y = 5$
$2(10) + 2y = 54$	$-10 + y = 5$
$20 + 2y = 54$	$y = 5 + 10$
$2y = 54 - 20$	$y = 15$
$2y = 34$	
$y = 17$	

$2(11) + 2y = 54 \longleftarrow x = 11 \longrightarrow -11 + y = 5$

$22 + 2y = 54$

$2y = 54 - 22$

$2y = 32$

$y = 16$

$y = 5 + 11$

$y = 16$

$2(12) + 2y = 54 \longleftarrow x = 12 \longrightarrow -12 + y = 5$

$24 + 2y = 54$

$2y = 54 - 24$

$2y = 30$

$y = 15$

$y = 5 + 12$

$y = 17$

$2(13) + 2y = 54 \longleftarrow x = 13 \longrightarrow -13 + y = 5$

$26 + 2y = 54$

$2y = 54 - 26$

$2y = 28$

$y = 14$

$y = 5 + 13$

$y = 18$

Here are the ordered pairs of coordinate points for the two linear equations.

$(9, 18), (10, 17). (11, 16), (12, 15), (13, 14)$ $(9, 14), (10, 15), (11, 16), (12,17), (13, 18)$

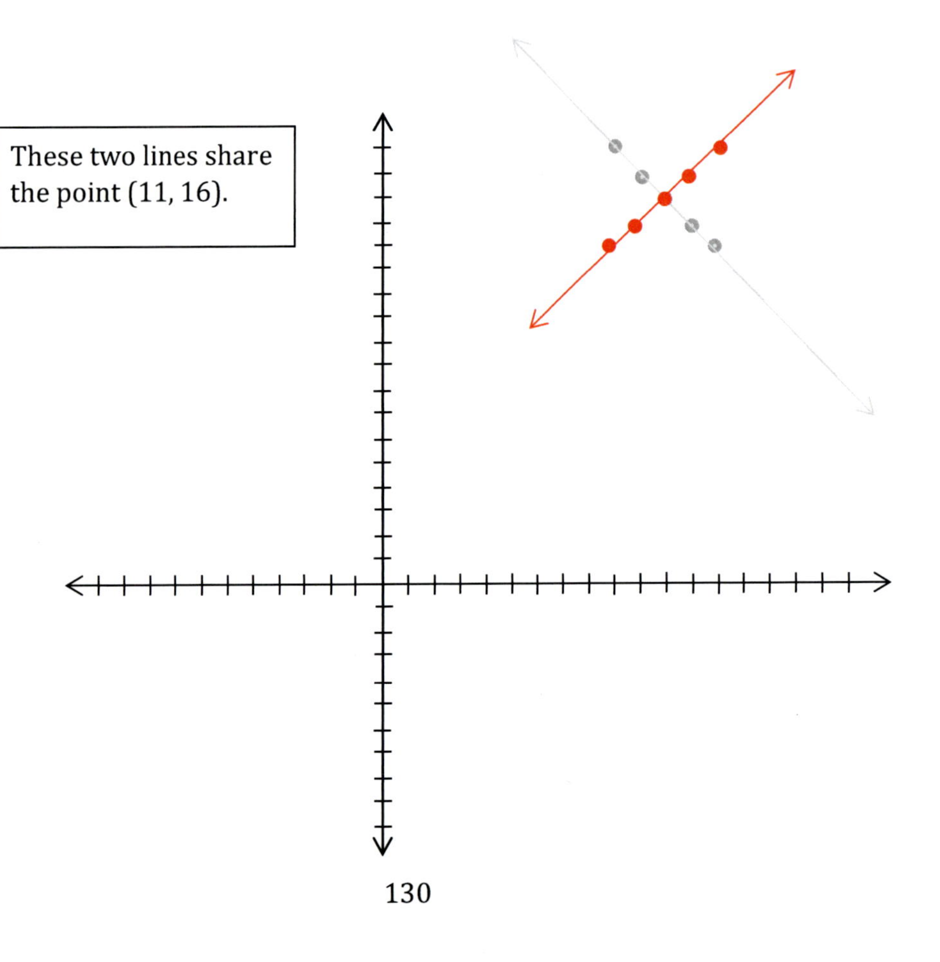

Do you understand what we are doing here? We solved a story problem by building two different linear equations. When we combined the two equations or used the substitution method, we were able to solve for x and y. The answers were $x = 11$ and $y = 16$ or (11, 16). Then we graphed those two linear equations by finding several other points. Once the lines were drawn, it was easy to see that the two lines intersected at (11, 16).

That's why building equations works! They will tell you the only two numbers that will solve your story problem. Pretty cool, huh?

Those last two examples were some of the easiest story problems with two variables, so let's try a more challenging one.

At a constant speed a pilot flew from town A to town B against the wind in 3 hours and returned in 2 hours. If the distance between A and B is 600 Kilometers, what were the rate of the wind and the speed of the plane?

Again, we are looking for two answers to solve one problem. We are trying to find the rate of the wind and the speed of the plane. Let's call the speed of the plane "x" and the rate of the wind "y." I'll draw a little picture too, so it's easier to visualize.

Here we show the plane going from Town A to Town B against the wind. Since the wind will slow the plane down, we have "plane speed minus wind" or $x - y$.

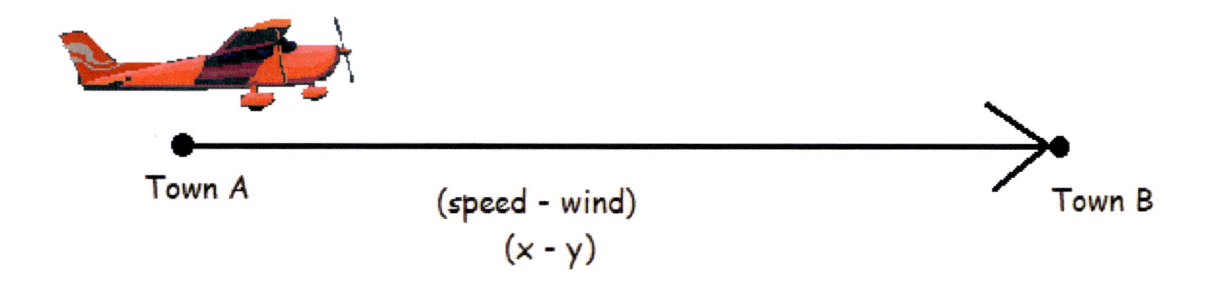

The next picture shows the plane returning to Town A. The wind is now pushing the plane faster, so we have "x PLUS y" as the rate of speed.

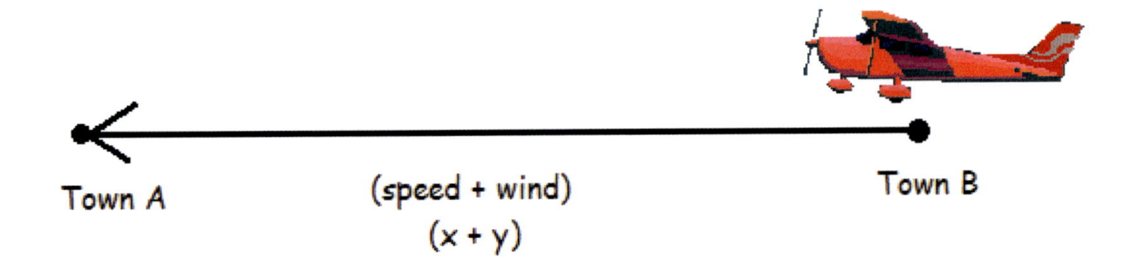

The first trip took 3 hours and the second trip took 2 hours. Both trips were 600 km in distance. Mmm…this sounds like a job for the Distance Formula, wouldn't you agree?

$$\frac{d}{r} = t$$

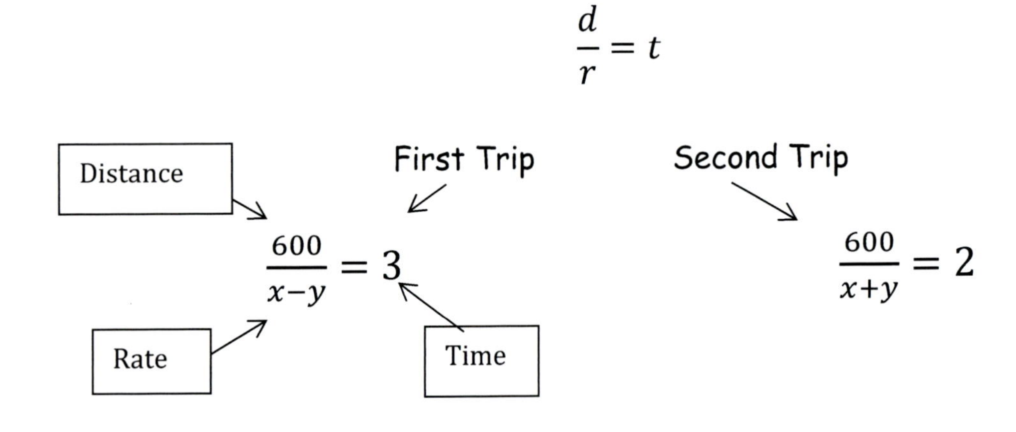

$$\frac{600}{x-y} = 3 \qquad \frac{600}{x+y} = 2$$

I'll solve the equation for the first trip, first.

$$\frac{600}{x-y} = 3$$

I can do the "el switch-a-roo" on this one.

$$\frac{600}{3} = x - y$$

$$200 = x - y$$

There is our first linear equation. Next, we will solve the equation for the second trip.

$$\frac{600}{x+y} = 2$$

Once again, it's time for the "el switch-a-roo."

$$\frac{600}{2} = x + y$$

$$300 = x + y$$

And now we have two linear equations, so let's combine them together.

$$200 = x - y$$
$$\underline{300 = x + y}$$

Well, look at that, we already have two terms that will cancel each other out; the y's. So let's get started.

$$200 = x - y$$
$$\underline{300 = x + y}$$
$$500 = 2x$$

$$250 = x$$

There we go, $x = 250$. Now let's solve for y. I'll bring down one of our equations for that.

$$300 = x + y$$

$$300 = 250 + y$$

$$300 - 250 = y$$

$$50 = y$$

Tada! The speed of the airplane (x) is 250 km/hr and the rate of the wind (y) is 50 km/hr.

If we plotted those linear equations onto a graph, we would see two lines intersect at (250, 50), but I think you get the idea.

If you able to easily solve these example problems on your own, continue on to the worksheet. But if you are confused, read this lesson again. It won't take long and it will make a lot more sense the second time around. Practice solving two-variable story problems, on the next worksheet.

WORKSHEET 13

Name ___________________________________ Date______________

1. The sum of two numbers is 43 and their difference is 9. Find the numbers.

2. Alvin is going to install some decorative tiles around a swimming pool. He will need to arrange 250 tiles using an assortment of plain white tiles and some expensive hand painted ones. The plain white tiles cost 50 cents each and the hand painted ones cost $3 each. His labor costs $250 and he is charging $800 for the whole job. How many of each tile should he use to stay within the budget?

3. A plane is flying against the wind from Town A to Town B. The two towns are 500 miles apart. It took the pilot 4 hours to get to Town B and 2 hours to get back to Town A. What were the speeds of the wind and the plane?

LESSON 14: QUADRATIC EQUATIONS

In Volume III of the *Learn Math Fast System*, you learned about linear equations. When you solved for x and y in a linear equation, you ended up with a set of coordinate points. If you plot several of these points on a graph and then connect the dots, you will draw a line. Get it? A **line**ar equation draws a **line.**

In Volume V of the *Learn Math Fast System*, you learned about quadratic equations. When you solved a quadratic equation, you always ended up with two possible answers for x, giving you two different coordinate points. When you plot several of those points on a graph, it will form a horseshoe shape. It might be upside down or right side up and it might be skinny or wide, but it will always draw a U-shape. Look at the parabola below.

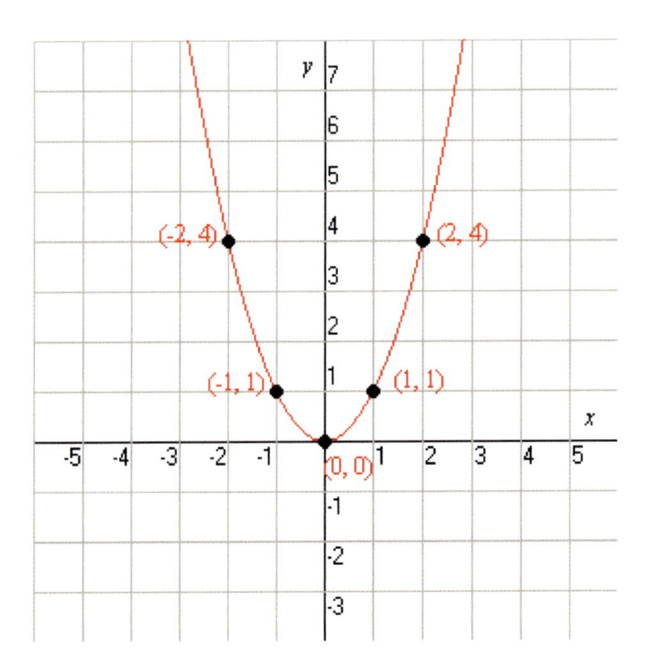

In the last lesson you built equations to solve story problems with two variables. You created two linear equations, added them together, and found the simultaneous solution to solve for x and y. When we plotted those linear equations onto a graph, we could visually see that our answers were also the point at which the two lines intersected.

Now we will go up to the next level. These story problems will produce a quadratic equation instead of a linear equation. Read the story problem below and then I will show you how to solve it.

> *We need to fence off a rectangular area in such a way that we get 60 square meters of space in which to park an RV. In order for the RV to fit, the length of the fence needs to be 4 meters longer than the width. How wide should the fence be in order to create this space?*

The way to solve this type of problem is to list all the known and unknown values (numbers), so we can create an equation to solve. One of the best ways to do that is to draw a picture. We know that our rectangular space needs to be 60 square meters in area, so let's start there.

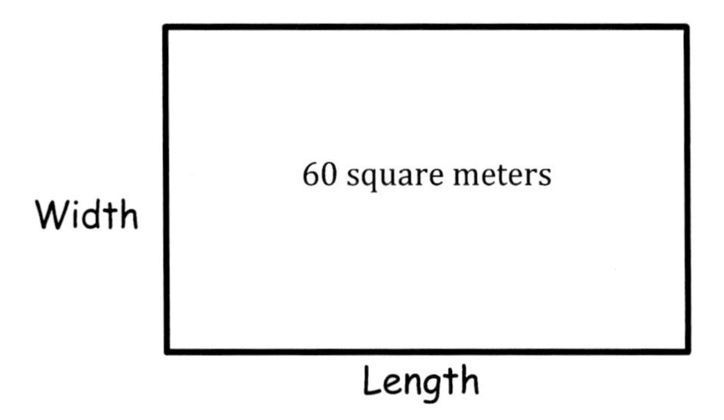

Do you recall how to find the area of a rectangle? You multiply length x width to get the area. Do we know either one of these dimensions? Well, we know that the length is 4 meters longer than the width, but we don't know the width. The width is our unknown number, so we will call the width, "x." Now look at our drawing.

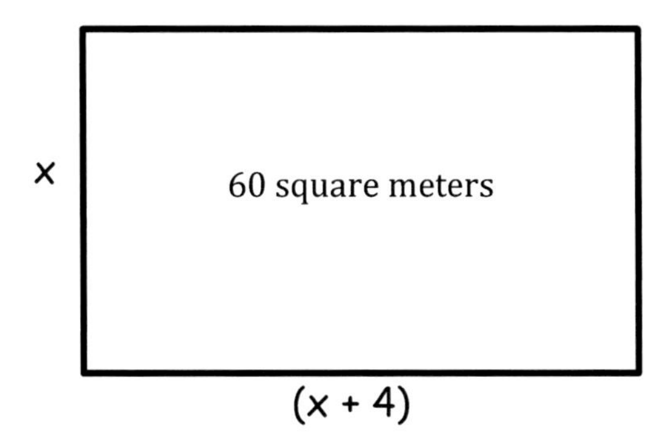

Since the width is our unknown, we are calling it x. We KNOW that the length is 4 meters longer than the width, so the length must be (x + 4). Now we know all of our "known and unknown values," so it is time to make an equation. The formula to find the area of a rectangle is "A = bh," so the equation below must be accurate. The formula uses base times height, which is the same thing as length times width.

$$x(x+4) = 60$$

Width x length = Area

All we have to do now is solve for x. Let's see, it looks like we need to use the Distributive Property of Multiplication on this one.

$$x(x+4) = 60$$

$$x^2 + 4x = 60$$

Now what? Well, this equation almost fits into the mold of a Standard Quadratic Equation. I'll just swing that 60 over to the other side and then it will be in the Standard Form.

$$x^2 + 4x - 60 = 0$$

There we go, now I can factor this polynomial to find some values for x. Do you remember how to factor a polynomial from Volume 5?

The first step is to make two sets of empty parentheses below our quadratic equation.

$$x^2 + 4x - 60 = 0$$

$$(\quad)(\quad) = 0$$

Can you figure out the rest from here? Give it a try on a separate piece of paper and then read on, to see if we got the same answers.

$$x^2 + 4x - 60 = 0$$

$$(\,x \quad\;)(\,x \quad\;) = 0$$

$$(\,x \quad 6\,)(\,x \quad 10) = 0$$

$$(\,x - 6\,)(\,x + 10) = 0$$

In order for this amount, times this amount, to equal zero, one of those amounts must also equal zero. That gives us two possible solutions for x.

$$x = 6 \;\; or \;\; x = \; -10$$

In this situation, we will have to use a little common sense. Which answer do you think makes the most sense? Do you think the width of the fence is 6 meters or negative 10 meters? Of course, x = 6 is the logical choice. Let's go back to our drawing and fill in x with 6.

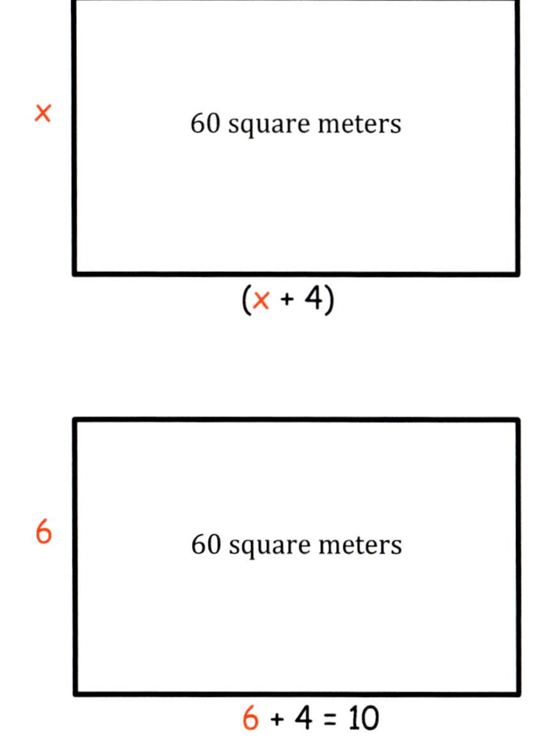

And now let's fill in our original equation with x = 6, to make sure it's right.

ORIGINAL EQUATION: $x(x+4)=60$

EQUATION WITH x = 6: $6(6+4)=60$

You could also solve this one like this:
$36+24=60$

$$6(10)=60$$
$$60=60$$

Yep, 60 does equal 60, so we got it right! The width of the fence should be 6 meters and the length is 10.

Let's try another one together before you try this on your own. This one will be a little more challenging.

Find the base of a triangle with an area of 6 square meters, if the altitude (height) of the triangle is 4 meters less than the base.

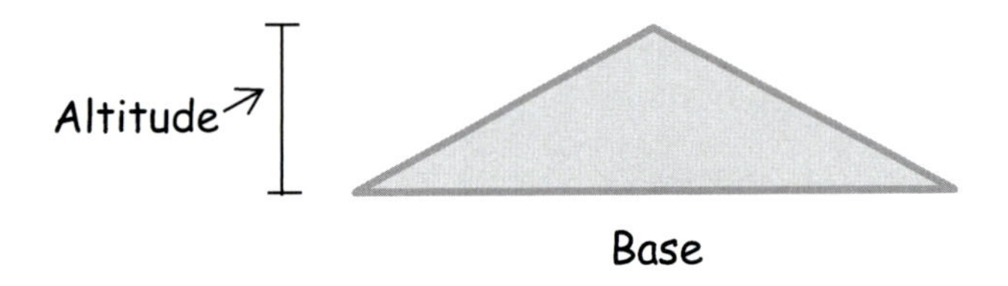

This problem deals with the area of a triangle. Do you know the "Area of a Triangle" formula? I'll remind you, in case you forgot.

$$A=\frac{1}{2}bh$$

This formula says, "Area equals one half of the base times the height." We need to fill in this formula with the information we have. We know the Area.

$$6\ square\ meters=\frac{1}{2}bh$$

We are looking for the base of the triangle; that is our unknown. We will call the base "x."

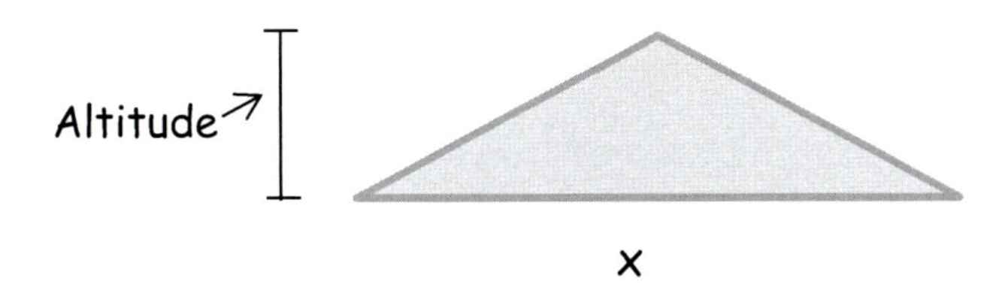

This story problem uses the word "altitude" instead of height, but they mean the same thing. Do you know the height or altitude of this triangle? The problem says it is 4 meters less than the base. How do you write that? If the base is x meters, then the height is "x - 4," right?

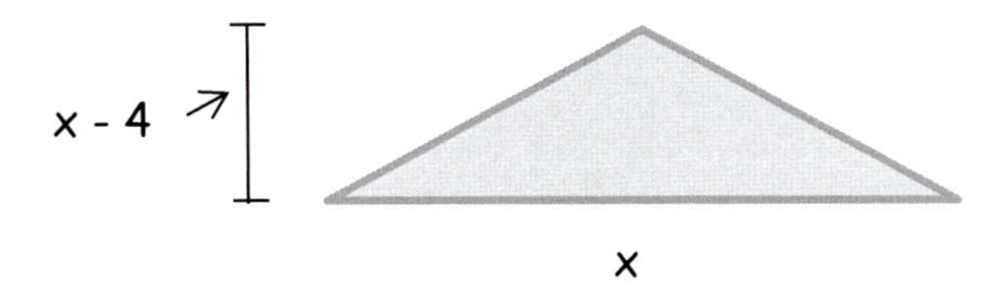

OK, now let's finish filling in our area formula.

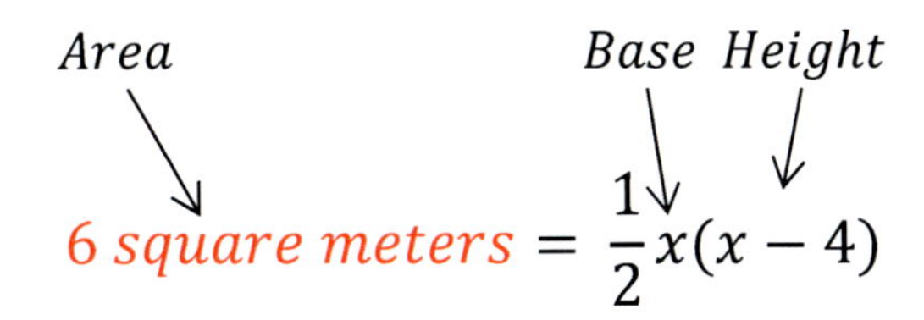

Let's do the math. Try to solve this one by yourself first. If you get stuck, study the work I've done below.

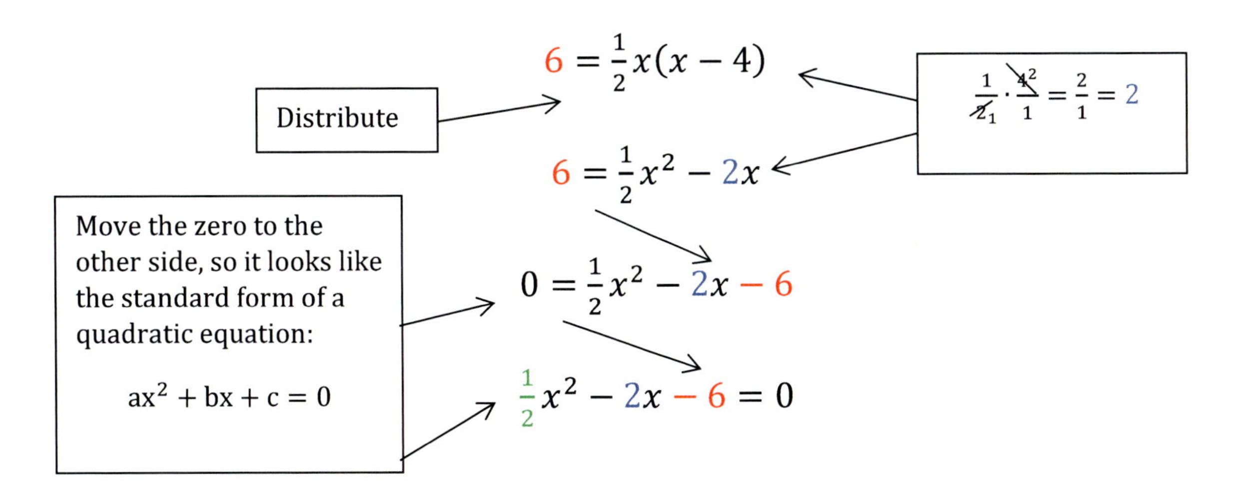

The "a" term, or the coefficient, of x^2 is $\frac{1}{2}$. When there is a number in place of "a," it is best to get rid of it. The way to get rid of that $\frac{1}{2}$ is to divide. Do you know why we need to divide? Since $\frac{1}{2}$ is multiplied by x^2, the opposite is division.

But we have to divide EACH term by $\frac{1}{2}$ in order for the equation to stay true. Here is the math to show each term being divided by $\frac{1}{2}$.

$$\frac{\frac{1}{2}x^2}{\frac{1}{2}} - \frac{2x}{\frac{1}{2}} - \frac{6}{\frac{1}{2}} = 0$$

I know that looks awful, but think back to your days of learning fractions and you'll see that it's not as difficult as it looks. Those are three different division problems. The first one is easy because we are just getting x^2 by itself. The second division problem is written in red below.

$$\frac{\frac{1}{2}x^2}{\frac{1}{2}} - \frac{2x}{\frac{1}{2}} - \frac{6}{\frac{1}{2}} = 0$$

I will rewrite that division problem, so it looks easier to solve.

$$\frac{2x}{1} \div \frac{1}{2} = \qquad \frac{2x}{1} \times \frac{2}{1} = \frac{4x}{1}$$

Flip and multiply

This is what we have so far:

$$x^2 - 4x - \frac{6}{\frac{1}{2}} = 0$$

Now let's solve the third division problem, written in blue. I'll rewrite it.

$$\frac{6}{1} \div \frac{1}{2} = \qquad \frac{6}{1} \times \frac{2}{1} = \frac{12}{1}$$

The third answer is 12. On a side note, when I see $\frac{6}{1} \div \frac{1}{2}$, I picture 6 sandwiches divided into or cut into halves. Now how many sandwiches do you have? I can see 12 small half sandwiches, do you? If your mind can't see that, just do the math. Now we have this equation.

$$x^2 \quad - \quad 4x \quad - \quad 12 = 0$$

Tada! We have a quadratic equation. Now let's solve for x. Do you know where to begin? We need to factor the quadratic.

$$x^2 \quad - \quad 4x \quad - \quad 12 = 0$$

$$(x \qquad)(x \qquad) = 0$$

Now we need two numbers that will equal -12 when multiplied and -4 when added. Or in fancy math terms, find two numbers with the product of -12 and the sum of -4.

So let's see…$-3 \times 4 = 12$, but those two numbers will never add up to -4. How about -2×6? That equals -12 too, but it adds up to $+4$. I'll just switch around the two signs, so it will still equal -12 and sum up to -4. Our two numbers are $2\ and - 6$.

$$(x + 2)(x - 6) = 0$$

In order for this equation to be true, one of the two sets of parentheses needs to equal 0, right? In that case, there are two possible answers for x.

$$x = -2 \quad or \quad x = 6$$

Again, we use a little common sense and decide that $x = 6$ makes the most sense. There is no way that our triangle is -2 meters tall.

I'll bring back our triangle, so we can replace x with our answer.

$x = 6$

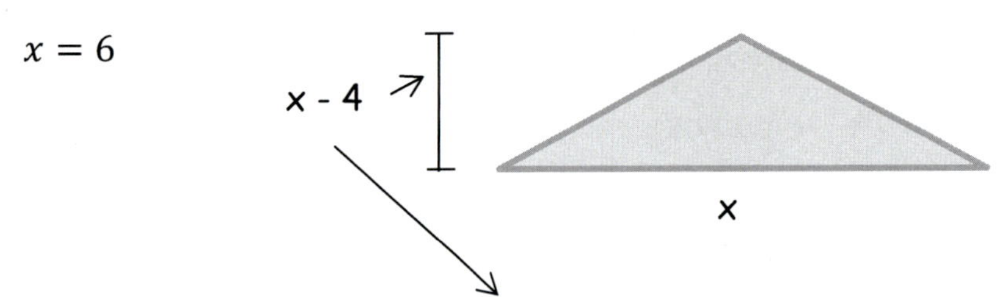

Height of the Triangle: $6 - 4 = 2\ meters$
Base of the Triangle: $6\ meters$

If this were a test, you would want to go back and reread the story problem to make sure you answer the right question. Our problem asked for the length of the base. The correct answer to this problem is 6 meters.

This next example is a common story problem used to teach quadratic equations. I will show you all the steps to solve it and then I will show you MY way. It is much faster and doesn't involve quadratics. But unfortunately, we will have to go through all the painful steps first, so you can learn how to solve a quadratic equation.

I need to make a box without a top. The box needs to be 3 centimeters deep with a total volume of 150 cm^3. The length of the box needs to be twice as long as the width of the box. The box maker would like to cut 3 cm squares out of the corners of a flat, rectangular piece of cardboard and then flip up the sides to create the box. What should be the size of the original piece of cardboard?

We are looking for the length and width of this piece of cardboard, before it is cut out. I will label the length of the original piece of cardboard as "x."

Once the box maker cuts out the 3cm x 3cm squares, the length of the box will be 6 cm shorter than x, right?

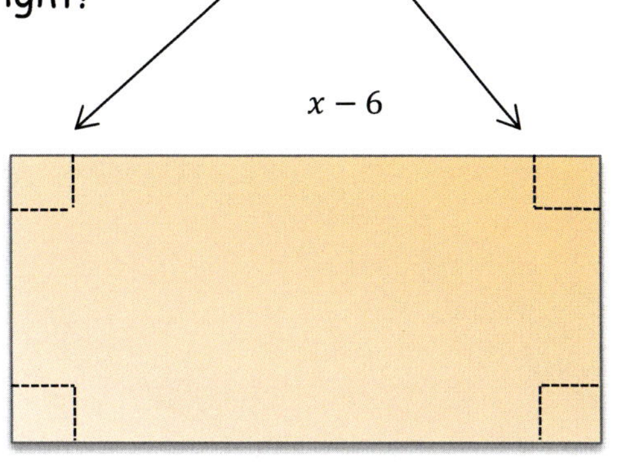

Now let's talk about the width of the box. The problem states that the length is twice as long as the width. So, wouldn't you agree that the width is HALF of the length? To find "half" of something, we divide by 2. The width of the box is the same as the length divided by 2.

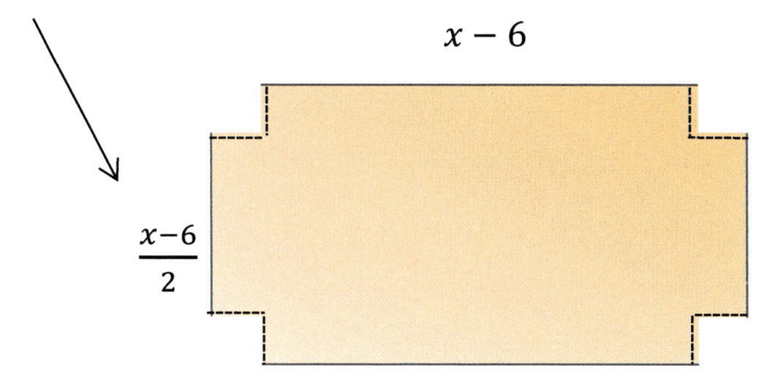

The problem also states that the volume of the box needs to be 150 cm^3. Do you know how to find the volume of a box? You multiply all 3 dimensions, right? The only dimension not written above is the depth. The box will be 3 cm deep.

I have multiplied the width times the length, times the depth of our box.

$$\frac{x-6}{2}cm \cdot x-6\ cm \cdot 3cm = 150cm^3$$

Take another look at that equation. I have rewritten it below, but this time I have colored each of the units red.

$$\frac{x-6}{2}cm \cdot x-6\ cm \cdot 3cm = 150cm^3$$

Notice that when we multiply cm x cm x cm, the answer will be cm^3. So for now, let's just drop the units. The units make the problem look too complicated.

$$\frac{x-6}{2} \cdot x-6 \cdot 3 = 150$$

That's better. Now focus your attention on this term. I want to get rid of that "divided by 2." The opposite is to "multiply by 2." If I multiply this term by 2, then I'll have to multiply the other side of the equal sign by 2 as well.

$$2\left(\frac{x-6}{2}\right) \cdot x-6 \cdot 3 = (150)2$$

Now our equation looks like this:

$$(x-6)(x-6)(3) = 300$$

Since this side of the equation is "multiplied by 3," let's get rid of it by dividing both sides by 3.

$$\frac{(x-6)(x-6)(3)}{3} = \frac{300}{3}$$

Now we have this:

$$(x-6)(x-6) = 100$$

Distribute (F.O.I.L.)

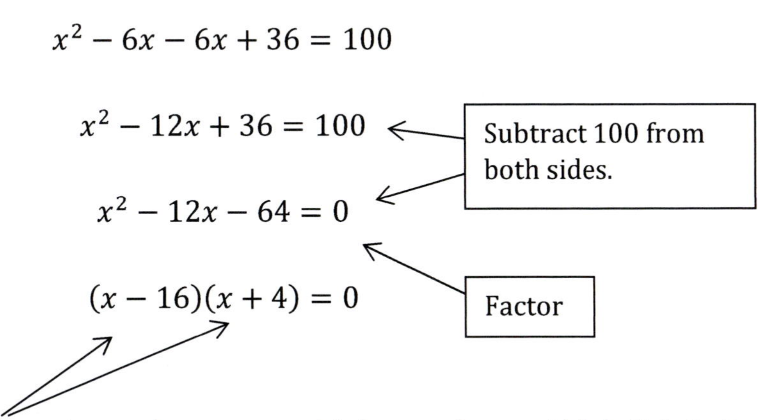

One of these two amounts must equal zero, so which one do you think it is? Do you think the length of the box is 16 centimeters or -4 centimeters? I say it's 16 cm.

Now that we know $x = 16$, can you solve for the width? On the last page, we said the width is $\frac{x-6}{2}$. Now can you tell me the width of the box?

$$\frac{16-6}{2} = \frac{10}{2} = 5$$

The BOX is 5 centimeters, but look back at the problem. We are to find the dimensions of the flat cardboard, not the box. The cardboard is 6 cm longer than the width of the box. The width is 11 centimeters and the length is 16 cm.

But there is an easier way! I know that all of the dashed lines are 3 cm long. And I know that the length of the BOX is twice as long as the width. This time, I will make "x" the width of the box, instead of the width of the cardboard.

We already know the volume is 150 cm^3 and the depth is 3 cm, so let's build an equation with that information.

$$2x \cdot x \cdot 3 = 150$$

$$6x^2 = 150$$

Divide both sides by 6

$$x^2 = 25$$

Get the square root of both sides

$$x = 5$$

The width of the box is 5 cm. The flat cardboard is 6 cm longer. The length of the box is twice as much as the width, so it is 10 cm. The flat cardboard is 6 cm longer than that, so it is 16 cm long.

We got the same answer either way. So, keep in mind that there is often more than one way to solve a problem. Always draw a picture, if you can, and then use some logic to create an equation. Also, be sure to go back to the question to make sure you are answering the question properly.

You might want to read this lesson again, before you try to complete the next worksheet.

WORKSHEET 14

Name ____ Date ____________

1. The area of a garden is 48 square meters. If the base is 2 meters longer than the height, how much fencing will be required to enclose it?

2. I need to order a box that is 4 inches deep with a volume of 320 in^3. The length needs to be 2 inches longer than the width. The factory would like to cut 4 inch squares out of the corners of a flat piece of cardboard and then flip up the sides to create the box. What should be the size of the flat piece of cardboard?

Worksheet 14 page 2

3. You need to fence off a rectangular shape next to a river. The area of the enclosure needs to be 96 square meters and the length should be 4 meters longer than the width. What should be the dimensions of the 3 fence sections?

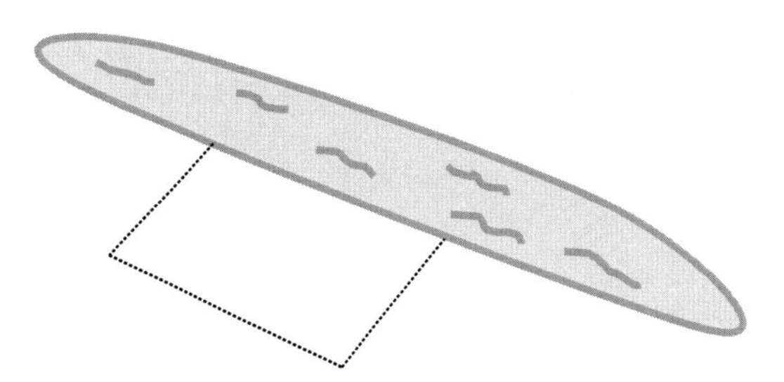

LESSON 15: GRAPHING A QUADRATIC EQUATION

Below is a difficult story problem. It involves a lot of math skills, but if you have read Volume 5, you have all the skills necessary to solve this one. Grab a piece of paper and write down this problem as we solve it together. It is a very long problem that involves A LOT of math, so maybe grab two pieces of paper.

> *A man is going to row a boat upstream for 2 miles and then turn around and row back downstream for 2 miles. He needs to do this in 1 hour and 20 minutes. If the rate of the stream is 2 miles per hour, how fast should he row?*

This one is challenging! First of all, we need to think about this logically. The problem mentions that the stream is moving at 2 miles per hour. So, when the man is rowing downstream at a particular speed, the stream is going to push him 2 miles per hour faster. And when the man is rowing upstream, against the current, he will be slowed down by 2 miles per hour.

We are trying to find out how fast he should row to get up and back in 1 hour and 20 minutes, keeping in mind that the stream is going to slow him down in one direction and speed him up in the opposite direction.

Since his "speed" is the unknown number, we are going to call it "x mph." When the man rows the boat upstream, he will be slowed down by 2 miles per hour, so his speed will be "x – 2 mph." And when he rows downstream, his speed is "x + 2 mph" because he gets sped up a bit by the current. So, let's write down everything we know, so far.

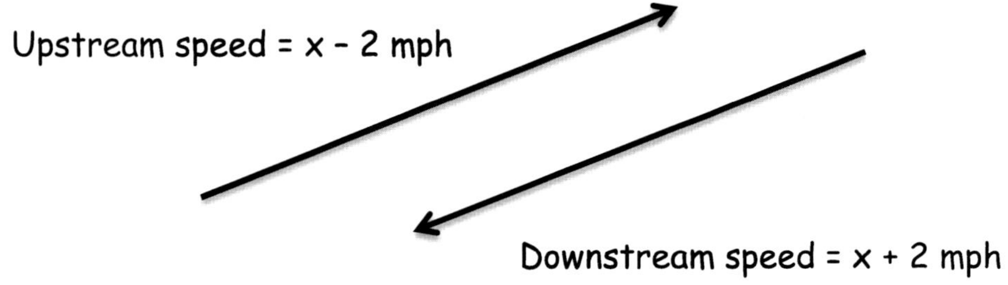

Total time = 1 hr. 20 min.
Distance = 2 miles up and 2 miles down

Now comes the tricky part. We have to make an equation, so we can solve for x. When you see words such as speed, distance, and time, you will know that it is time to use our "Distance Formula."

$$\frac{d}{r} = t$$

We need to replace these variables with the values from our story problem. Mmm…Let's think about this…the man traveled 2 miles at a rate of "x - 2" and then he traveled another 2 miles at a rate of "x + 2." We want him to complete this entire trip in 1 hour and 20 minutes, so the time it takes to go upstream PLUS the time it takes to go downstream, needs to equal 1 hour and 20 minutes.

OK, well if we have two different rates and two different distances, let's do this:

Upstream time + Downstream time

$$\frac{d}{r} + \frac{d}{r} = t$$

Total time

NOW we can replace the variables with our values.

$$\frac{2\ miles}{x - 2} + \frac{2\ miles}{x + 2} = 1\ hour\ 20\ min.$$

Mmm…before we can continue, I'm going to have to take issue with our units of time. We have hours AND minutes here. That is going to mess up our math. We need our units of time to be in one unit, either minutes or hours, but not both. Since we are also dealing with "miles per hour," let's turn "1 hour and 20 minutes" into just hours.

So, let's see…20 minutes is what fraction of 60 minutes? That can be solved with algebra too!

$$20 = x \cdot 60\ minutes$$

$$20 = 60x$$

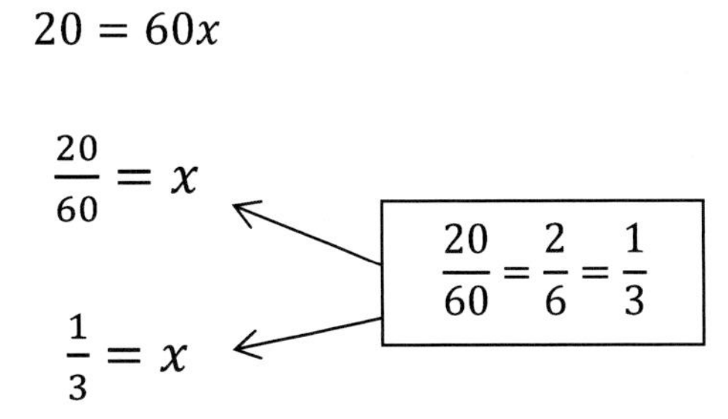

So, how long is the boat trip now? It is $1\frac{1}{3}$ hours. But mixed numbers are a bit messy in algebra, so let's convert that time into an improper fraction.

$$1\frac{1}{3} = \frac{4}{3}$$

Now I'll rewrite our equation with only one unit of time. Write this on your piece of paper too. This is the equation that we need to solve.

$$\frac{2}{x-2} + \frac{2}{x+2} = \frac{4}{3}$$

Look at this equation as two fractions being added together, just like these two fractions are being added together.

$$\frac{10}{2} + \frac{6}{3} = \frac{21}{3}$$

And how do you add fractions together? You need to find a common denominator, right? Well then, that's what we need to do with our equation; find a common denominator.

$$\frac{2}{x-2} + \frac{2}{x+2} = \frac{4}{3}$$

Can you guess how to do that? Do you remember what to do when you can't find a common denominator? You multiply the denominators together and use that number. For example, how would you find the common denominator in this problem?

$$\frac{4}{9} + \frac{1}{5} =$$

The least common denominator for these two fractions is 45. I figured that out by multiplying the two denominators. Is that what you did? I'll rewrite the equation with the common denominators and then add them together.

$$\frac{4}{9} + \frac{1}{5} = \quad \frac{20}{45} + \frac{9}{45} = \frac{29}{45}$$

Let's do the same thing with our equation; multiply the denominators together. But notice that our equation has three denominators. The fraction in the answer has to have the same denominator as well.

$$\frac{2}{x-2} + \frac{2}{x+2} = \frac{4}{3}$$

So, we have to multiply ALL THREE denominators together. That will make the Least Common Denominator in this equation $(3)(x-2)(x+2)$.

And you know the rule about getting a common denominator, don't you? Whatever you do to the denominator, you must do to the numerator. Let's focus on just the first fraction in our equation for now. It is written in blue below.

$$\frac{2}{x-2} + \frac{2}{x+2} = \frac{4}{3}$$

I will add our common denominator below.

$$\frac{2}{(x-2)(x+2)(3)}$$

Remember, whatever we did to the denominator, we must do to the numerator. What did we do to the denominator? We multiplied it by $(x+2)(3)$, so let's do that to the numerator too.

$$\frac{2(x+2)(3)}{(x-2)(x+2)(3)}$$

Now we can do the same thing to the other two fractions in our equation. Write this all out onto your piece of paper.

$$\frac{2(x+2)(3)}{(x-2)(x+2)(3)}+\frac{2(x-2)(3)}{(x+2)(x-2)(3)}=\frac{4(x-2)(x+2)}{3(x-2)(x+2)}$$

The original equation is written in blue. This is looking pretty intense. Luckily there is a short cut to this type of problem. To help you understand this short cut, I'm going to bring back an earlier example.

$$\frac{10}{2}+\frac{6}{3}=\frac{21}{3}$$

There are a few different ways to add $\frac{10}{2}+\frac{6}{3}$ together to see if they really do equal $\frac{21}{3}$. One way is to look closely at the equation above and realize that each one of those fractions is actually a whole number. Follow the arrows below.

$$\frac{10}{2}+\frac{6}{3}=\frac{21}{3}$$

$$\frac{10}{2}=5 \qquad \frac{6}{3}=2 \qquad \frac{21}{3}=7$$

$$5+2=7$$

That is the fastest way to add those fractions together. Another way would be to find a common denominator. I've done that math below.

$$\frac{10}{2}+\frac{6}{3}=\frac{21}{3} \qquad \frac{30}{6}+\frac{12}{6}=\frac{42}{6} \textit{ reduces to } 7$$

That way takes longer, but you get the same answer. There is one more way to see if this equation is true. It isn't the fastest way, but when you get to an intense equation like the one we are working on, it will be faster and I'll show you why.

Take one more look at our example problem again and watch this trick.

$$\frac{10}{2}+\frac{6}{3}=\frac{21}{3}$$

If I multiply all three of those denominators together ($2 \cdot 3 \cdot 3$), I get the number 18. I can take that number and multiply it by each term. This will turn each fraction into a whole number. Of course, the other methods are faster, but they won't work with our row boat equation, so we will go through each painful step.

$$\frac{18}{1}\cdot\frac{10}{2}+\frac{6}{3}\cdot\frac{18}{1}=\frac{21}{3}\cdot\frac{18}{1}$$

This involves multiplying fractions. When you multiply fractions, you are allowed to cross cancel. Write the equation above onto your paper and cross cancel where ever possible.

$$\frac{\overset{9}{\cancel{18}}}{1}\cdot\frac{10}{\underset{1}{\cancel{2}}}+\frac{6}{\underset{1}{\cancel{3}}}\cdot\frac{\overset{6}{\cancel{18}}}{1}=\frac{21}{\underset{1}{\cancel{3}}}\cdot\frac{\overset{6}{\cancel{18}}}{1}$$

$$\frac{90}{1}+\frac{36}{1}=\frac{126}{1}$$

$$or\ 90+36=126$$

OK, now that you have seen how that works, let's apply that little trick to our row boat equation. I'll multiply all three denominators together and then multiply that answer (written in red on the next page) by each term in this equation.

$$\frac{2}{(x-2)}+\frac{2}{(x+2)}=\frac{4}{3}$$

$$\frac{3(x-2)(x+2)}{1} \cdot \frac{2}{(x-2)} + \frac{2}{(x+2)} \cdot \frac{3(x-2)(x+2)}{1} = \frac{4}{3} \cdot \frac{3(x-2)(x+2)}{1}$$

Next, we can cross cancel some of the factors. Remember the rule to canceling: Factors are Fine, but Terms are Trouble. That means that we can cancel FACTORS, but we cannot cancel TERMS. Our equation is full of nothing but factors, so no worries here. I have cross canceled the identical factors in each multiplication problem.

$$\frac{3\cancel{(x-2)}^{1}(x+2)}{1} \cdot \frac{2}{\cancel{(x-2)}_{1}} + \frac{2}{\cancel{(x+2)}_{1}} \cdot \frac{3(x-2)\cancel{(x+2)}^{1}}{1} = \frac{4}{\cancel{3}_{1}} \cdot \frac{\cancel{3}^{1}(x-2)(x+2)}{1}$$

Here is what we have left:

$$\underbrace{\frac{3(x+2)}{1} \cdot \frac{2}{1}} + \underbrace{\frac{2}{1} \cdot \frac{3(x-2)}{1}} = \underbrace{\frac{4}{1} \cdot \frac{(x-2)(x+2)}{1}}$$

$$3(x+2) \cdot 2 \quad + 2(3)(x-2) = \quad 4(x-2)(x+2)$$

I'll rearrange and simplify this expression.

$$6(x+2) + 6(x-2) = (4x-8)(x+2)$$

I'll use the Distributive Property on these terms and the "FOIL" method here.

$$6x + 12 + 6x - 12 = 4x^2 + 8x - 8x - 16$$

Combine the like terms.

$$12x = 4x^2 - 16$$

Rearrange these terms, so they are in the standard form of a quadratic equation.

$$4x^2 - 12x - 16 = 0$$

Standard Form of a Quadratic Equation
$\mathrm{ax^2 + bx + c = 0}$

Mmm…there is a number in the "a" term. We want to get rid of that, otherwise, this is going to be way too hard to solve. Let's divide each term by 4.

$$\frac{4x^2}{4} - \frac{12x}{4} - \frac{16}{4} = 0$$

$$x^2 - 3x - 4 = 0$$

Factor the quadratic equation

$$(x \quad\ )(x \quad\ )$$

Do you know what to do next? We are trying to find the two factors that were multiplied together to get $x^2 - 3x - 4 = 0$. In other words, we want to "un-foil" the quadratic to get back to the two sets of parentheses. I already filled in the two x's. Next, we need to find two numbers that will equal -4 when multiplied and -3 when added.

In order to get negative product, the two factors must have different signs. What two numbers equal 4 when multiplied? It's either 2 x 2 or 4 x 1. Well, let's see 2 and 2 will never equal -3 when added, but 4 and 1 will.

$$(x \quad 4)(x \quad 1)$$

If I make the 4 negative and the 1 positive, they will equal -3 when added and -4 when multiplied - Perfect!

$$(x - 4)(x + 1) = 0$$

OK, we are almost done. In order for this equation to equal 0, one of these two factors must also equal zero, right? If $x = 4$, then that equation would equal 0. Or if $x = -1$, then it would also equal 0. So those are the two possible answers.

$$x = 4 \; or \; x = -1$$

In a quadratic equation, you will always end up with two possible answers for x. You need to think about the answers logically and decide which one makes the most sense. We are trying to find the speed of the boat, so I doubt that -1 is the answer. I say the man should row his boat at 4 miles per hour in order to get up and back in 1 hour and 20 minutes.

Let's fill in our equation with 4 miles per hour, to see if it is right.

$$\frac{2}{(x-2)}+\frac{2}{(x+2)}=\frac{4}{3}$$

$$\frac{2}{(4-2)}+\frac{2}{(4+2)}=\frac{4}{3}$$

$$\frac{2}{2}+\frac{2}{6}=\frac{4}{3}$$

Ooo, let's try that cool, new trick we just learned. I'll multiply all three denominators together and then multiply each fraction by that number, cross canceling when I can.

$$2\times 6\times 3=36$$

$$\frac{\overset{18}{\cancel{36}}}{1}\cdot\frac{2}{\underset{1}{\cancel{2}}}+\frac{\overset{6}{\cancel{36}}}{1}\cdot\frac{2}{\underset{1}{\cancel{6}}}=\frac{4}{\underset{1}{\cancel{3}}}\cdot\frac{\overset{12}{\cancel{36}}}{1}$$

$$18\cdot 2+6\cdot 2=4\cdot 12$$

$$36+12=48$$

Yep, $36+12$ does equal 48, so we got it right! The man should row at 4 miles per hour in order to get upstream and back in 1 hour and 20 minutes.

OK, now that we know the answer to our story problem, let me show you something very interesting. We ended up with this quadratic equation $x^2-3x-4=0$, right?

When our quadratic equation was set to equal 0, the values we found for "x" were $x = 4 \; and \; x = -1$. If you were to plot these points onto a graph, they would both have an ordinate point (remember, that's the "y" point) of 0.

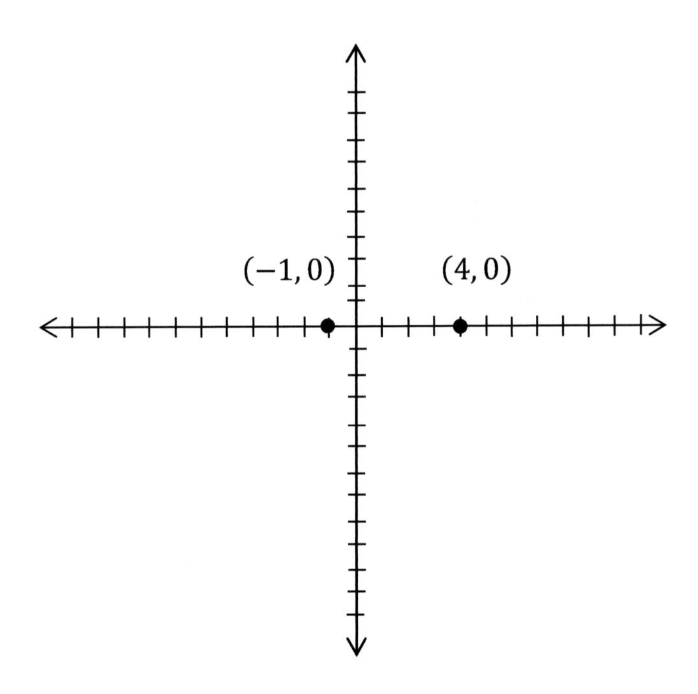

I'm sure you are tired of this row boat story problem by now, but you really need to see the whole equation plotted on a graph because there is more to learn from our equation. In order to do that, we will need some more points. We could assign some values to "x" and solve for "y," but this is QUADRATICS, my friend. This isn't for little kids, so let's make it look more sophisticated with functions!

Do you recall learning about functions? Do you remember this symbol?

$$f(x)$$

Look at this quadratic equation again.

$$x^2 - 3x - 4 = 0$$

Since that quadratic ended with "equals zero," it is an EQUATION. It is purposely set to "equal zero" so we can find the "x" coordinates when "y" equals zero.

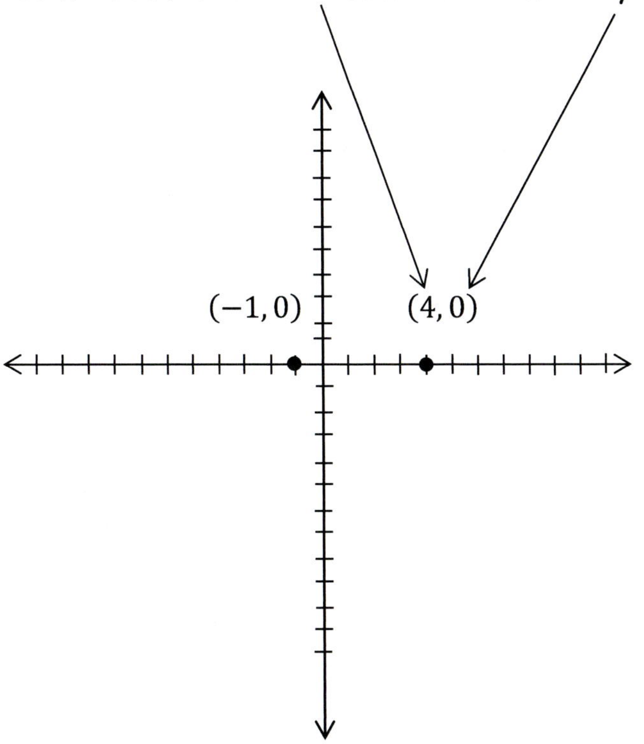

This is a quadratic function.

$$f(x) = x^2 - 3x - 4$$

We can use it to find more points on our graph. I will fill in the red "x" in "$f(x)$" with a number; I choose 3.

$$f(3) = x^2 - 3x - 4$$

When you see this, you are to replace every "x" with the number 3. Then we can solve the function.

$$f(3) = x^2 - 3x - 4$$

$$f(3) = 3^2 - 3 \cdot 3 - 4$$

$$f(3) = 9 - 9 - 4$$

$$f(3) = -4$$

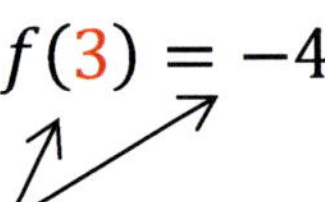

Those are the next two x, y coordinates for our graph. The math above is saying when you plug the number 3 into our function, the output, y, will be -4. Let's plot (3, -4) onto our graph.

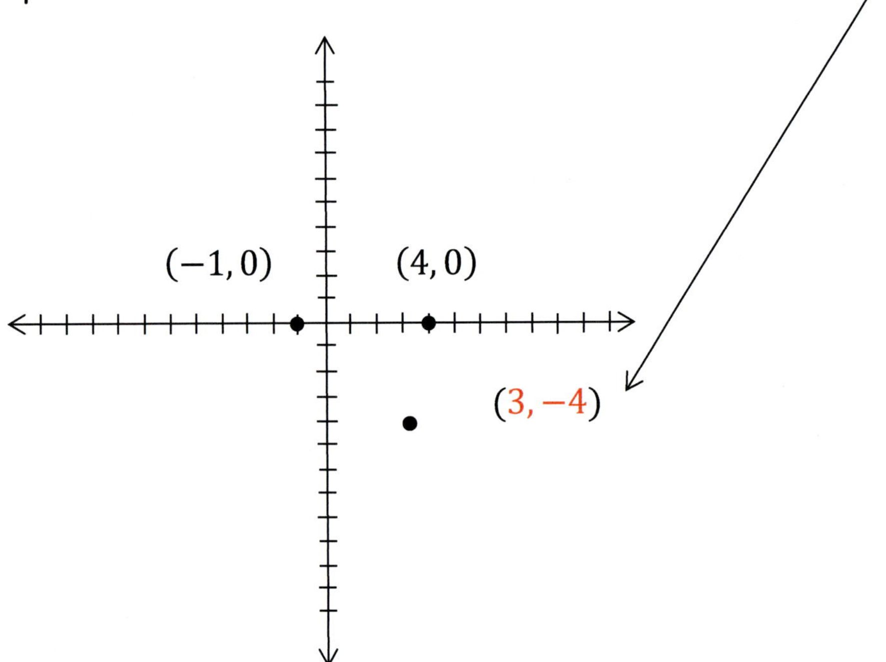

A quadratic function will always draw a parabola on a graph. So far, our dots do not form that horseshoe shape, so we need to find more points.

I'll draw a table, so we can see all the points we create as $f(x)$ changes.

$$f(x) = x^2 - 3x - 4$$

$f(-2)$	$f(-1)$	$f(0)$	$f(1)$	$f(2)$	$f(5)$

This table is saying, "Replace x with the number in the parentheses." Let's get started, beginning with $f(-2)$.

$$f(-2) = -2^2 - (3 \cdot -2) - 4$$

$$f(-2) = 4 - (-6) - 4$$

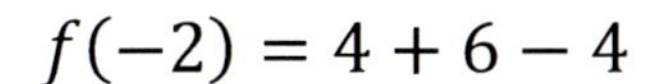

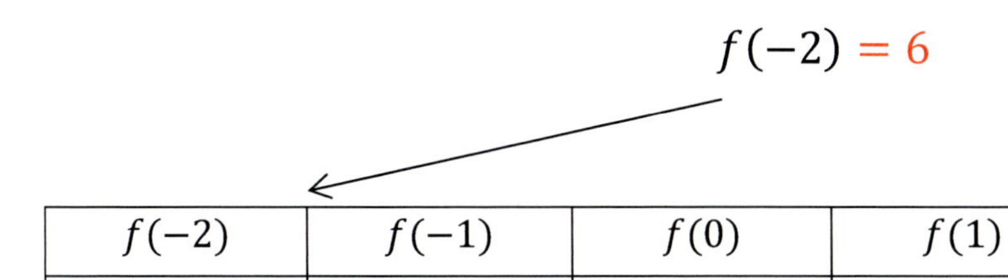

$f(-2)$	$f(-1)$	$f(0)$	$f(1)$	$f(2)$	$f(5)$
6					

Get out a piece of paper and solve the rest of these by yourself. Then compare your answers to mine below.

$f(-1) = -1^2 - (3 \cdot -1) - 4$
$f(-1) = 1 - (-3) - 4$
$f(-1) = 0$

$f(0) = 0^2 - (3 \cdot 0) - 4$
$f(0) = 0 - 0 - 4$
$f(0) = -4$

$f(1) = 1^2 - (3 \cdot 1) - 4$
$f(1) = 1 - 3 - 4$
$f(1) = -6$

$f(2) = 2^2 - (3 \cdot 2) - 4$
$f(2) = 4 - (6) - 4$
$f(2) = -6$

$f(5) = 5^2 - (3 \cdot 5) - 4$
$f(5) = 25 - 15 - 4$
$f(5) = 6$

Did you get the same answers? I added my answers to the table. Now I can plot the points on our graph.

$f(-2)$	$f(-1)$	$f(0)$	$f(1)$	$f(2)$	$f(5)$
6	0	-4	-6	-6	6

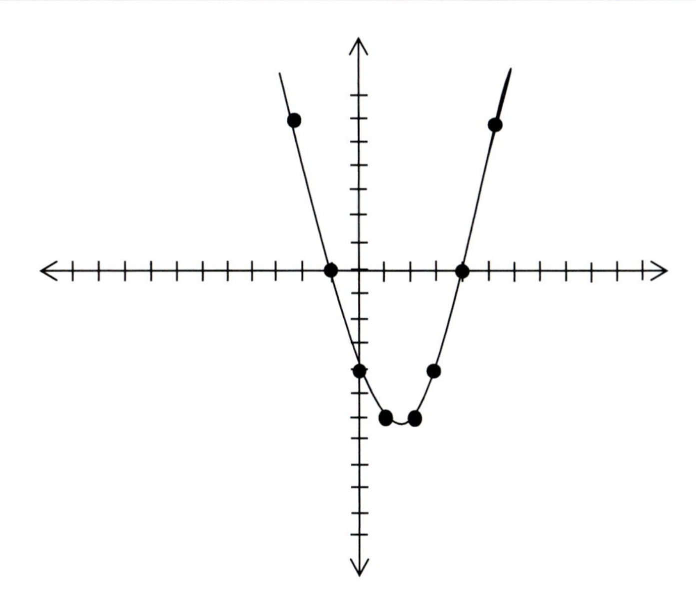

Look at that! Our points drew out a horseshoe shape. That U-shape is called a parabola (pa-rab-ola). Now think back to our story problem. We wanted to know the ideal speed for the boater to row in order to complete the trip in 1 hour and 20 minutes. When we set our quadratic equation to equal zero, we forced that equation to tell us the value of "x" when "y = zero." The answer, 4, was the only number that would make our boating equation come true.

We are going to quickly go through one more of these row boat questions because sometimes they don't work out as easily as the last one did. Here is our next example. Try to solve it on your own first. If you get stuck, read my solution.

A woman is going to row a boat upstream at a steady pace for 3 miles. The rate of the current is 3 mph. She needs to row up the stream and then back down in 1 hour and 20 minutes. How fast should she row?

This equation is built the same way as the last one.

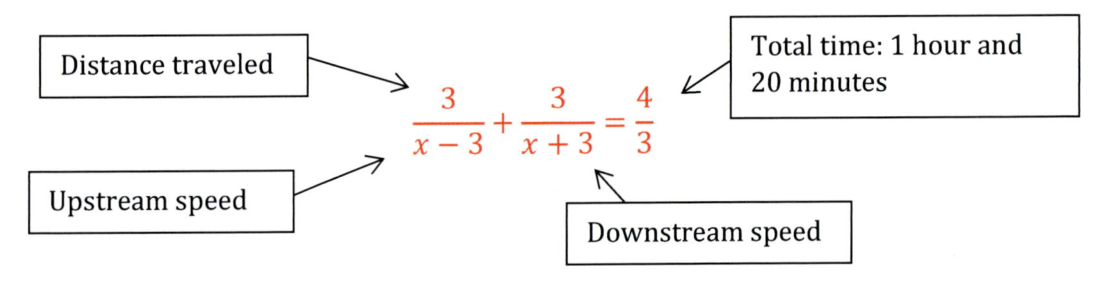

The next step is to get a common denominator. Do you remember how to do that? Multiply the 3 denominators together. Then, multiply that number by each numerator to get back to whole numbers. I've done that below.

Common Denominator

$(x-3)(x+3)(3)$

$$\frac{3}{\cancel{x-3}_{1}}\cdot\frac{\cancel{(x-3)}^{1}(x+3)(3)}{1}+\frac{3}{\cancel{x+3}_{1}}\cdot\frac{(x-3)\cancel{(x+3)}^{1}(3)}{1}=\frac{4}{\cancel{3}_{1}}\cdot\frac{(x-3)(x+3)\cancel{(3)}^{1}}{1}$$

$$[3(x+3)\cdot 3]+[3(x-3)\cdot 3]=4(x-3)(x+3)$$

I will simplify that equation for you.

$$[3(x+3)\cdot 3] + [3(x-3)\cdot 3] = 4(x-3)(x+3)$$

$$9(x+3) + 9(x-3) = 4(x^2-9)$$

$$9x + 27 + 9x - 27 = 4x^2 - 36$$

Simplify again

$$18x = 4x^2 - 36$$

Swing the $18x$ over to the other side and change the sign

$$0 = 4x^2 - 18x - 36$$

Let's stop here for a moment and focus your attention on the "a" term, $4x^2$ in the equation above. Do you know why I keep calling it the "a" term? Do you recall the Standard Form of a Quadratic Equation? It looks like this:

$$ax^2 + bx + c = 0$$

This term is the "a" term. The variable "a" is the coefficient of x^2. Sometimes the coefficient will be the number 1, but other times you won't be that lucky and this is one of those times.

The best option would be to divide each term by that coefficient to make it go away, but sometimes, even that won't work. We could divide each term below by the number 4, but 18 isn't divisible by 4.

$$0 = 4x^2 - 18x - 36$$

If we did that, we would end up with this equation:

$$0 = x^2 - \frac{18x}{4} - 9$$

I've tried to solve that one, but it's a real brain twister, so let's try something else.

$$0 = 4x^2 - 18x - 36$$

Each one of the numbers, in the equation above, is an EVEN number. We know that even numbers are always divisible by 2, so let's divide each term by 2 and see what we get.

$$0 = \frac{4x^2}{2} - \frac{18x}{2} - \frac{36}{2}$$

$$2x^2 - 9x - 18 = 0$$

OK, I think I can work with this. There is just one extra step or bit of skill that I will have to show you. When I factor this quadratic equation into two sets of parentheses, the first term will be a $2x$.

$$(2x \quad)(x \quad)$$

So now, when you are trying to come up with two numbers that equal -9x when added and -18 when multiplied, you will have to keep in mind that this 2x is going to impact the last term.

$$(2x + 3)(x - 6) = 0$$

Think about the "FOIL" trick. The second step in the FOIL trick is to multiply the two "Outer" terms together. So, as I'm trying to come up with two numbers to go inside of the parentheses, I have to keep thinking about how that 2x is going to change things.

$$2x^2 - 9x - 18 = 0$$

$$(2x \quad)(x \quad)$$

The logical choice would be to put a 6 and 3 in here, but what signs would you use? If they are both negative, they won't equal -18 when multiplied. And if they are opposite signs, will they equal -9 when added? That's when you have to take that

2x into consideration. That 2x is going to be multiplied by the last term, before we add the center terms together. Let me show you what I mean.

$$2x^2 - 9x - 18 = 0$$

$$(2x + 3)(x - 6) = 0$$

I'll use the FOIL method to multiply these two factors.

$$(2x + 3)(x - 6) = 0$$

$$2x^2 \underbrace{- 12x + 3x}_{} - 18$$

Would you look at that? When I add these two center terms together, they equal -9x, so those numbers will work! Do you know what to do now?

$$(2x + 3)(x - 6) = 0$$

One of these two expressions must equal 0, in order for this equation to be true.

$$2x + 3 = 0 \quad or \quad x - 6 = 0$$

Solve for x.

$$2x = -3$$

$$x = -\frac{3}{2} \qquad or \quad x = 6$$

The woman should row at a rate of 6 miles per hour.

The point I'm trying to make is that even though it is best to get rid of the coefficient in the "a" term, sometimes that won't be the best option. You will have to logically come up with the best option.

Now for the bad news...I think you should read this entire lesson over again. I know, I know, and I'm sorry, but you really should. But this time, try to do all the work yourself before you read my work. THEN, when you feel like you know these problems inside and out, complete them from beginning to end in front of someone else. Explain each step to your audience as you impress them with your incredible math skills. Believe me, if you can complete these problems in front of someone, they will think (know) you are very smart!

Well, were they impressed? Now try some on your on by completing the next worksheet.

WORKSHEET 15

Name ______________________________ Date ________________

1. A man is going to row a boat up a stream for 4 miles. The rate of the current is 2 mph. He needs to row up the stream and then back down in 2 hours and 40 minutes. How fast should he row?

2. In the last problem, you ended up with an equation that was in the standard for of a quadratic equation. Write that equation here:

   ____

   Use that equation to fill in the table below and then plot those points onto the graph on the next page.

$f(-2)$	$f(-1)$	$f(0)$	$f(1)$	$f(2)$	$f(3)$	$f(4)$	$f(5)$

Worksheet 15 page 2

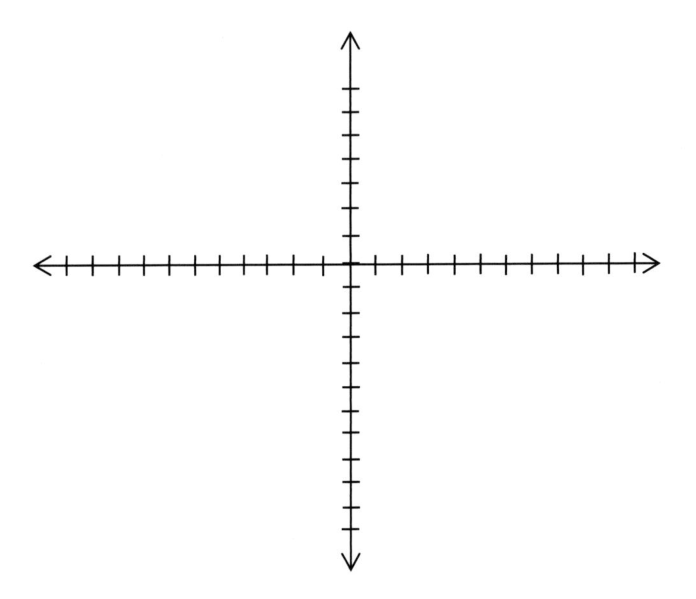

FINAL TEST

Name ______________________________________ Date ____________

1. A race car driver is trying to beat his opponent's record. His opponent drove 200 miles in 1 hour and 10 minutes. If the race car driver travels at an average speed of 175 miles per hour, will he beat his opponent's record?

2. Esther jogged at 4 mile per hour for 2 hours and 45 minutes. How far did she get?

3. Teresa and Jessica are sewing some aprons. Teresa has been sewing for many years and she can complete 3 aprons in just 1 hour. Jessica is new at this job and can only sew 1 apron in 2 hours. If the two of them work together for 8 hours, how many aprons will they make?

4. Steve, Mike, and Eric worked for 6 hours and 40 minutes cutting grass. After the first hour, Steve had mowed $\frac{3}{4}$ of the first lawn. Mike had mowed $\frac{1}{2}$ of the second lawn and Eric had mowed $\frac{1}{4}$ of the third lawn. If they keep up this same pace for 6 hours and 40 minutes, how many lawns will they mow?

Final Test page 2

5. The two triangles below are congruent. Solve for x.

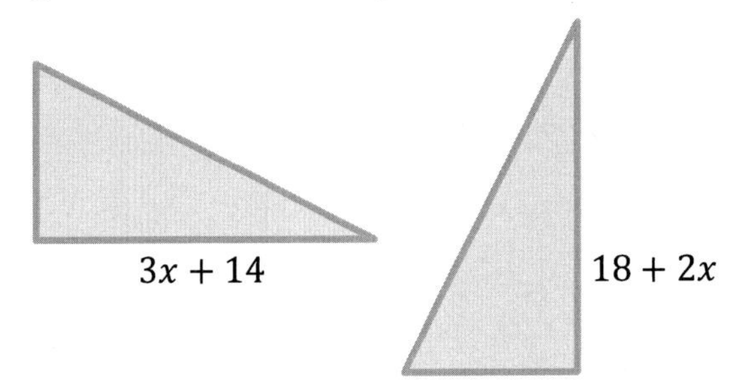

6. What is the area of the shaded triangle below?

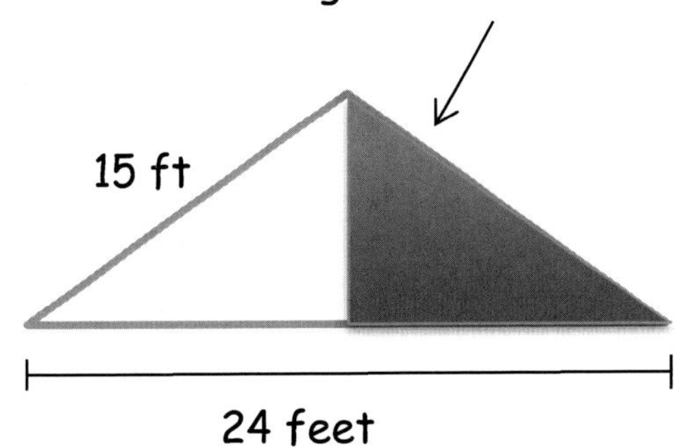

Final Test page 3

7. A builder needs to re paint the green triangular portion of this house a different color. He is told that the roof has a pitch of 6:12 and he knows that the blue board is 32 feet long. Each can of paint will cover 140 square feet. How many cans of paint will he need?

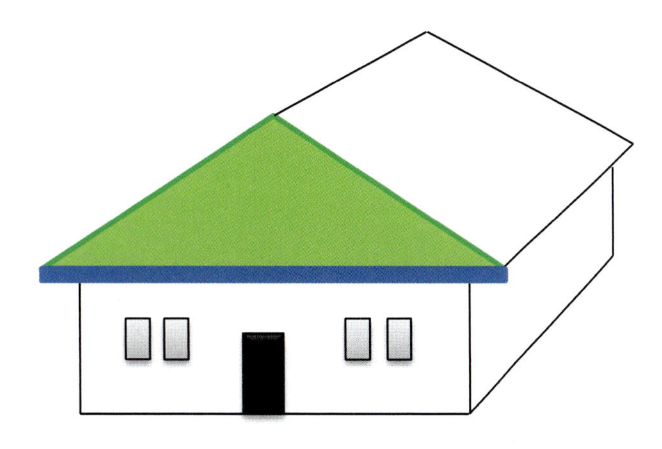

8. What is the probability of rolling an even number on a regular 6-sided die?

Final Test page 4

9. In a carton of a dozen eggs, 3 of the eggs are broken. You randomly select 2 eggs to cook. What is the probability of selecting 2 broken eggs?

10. A bowl contains 5 girl names and 7 boy names. What is the probability of randomly selecting two boy names or two girl names?

Solve the inequalities below.

11. $3(x+2) < 5(x-4)$

12. $\frac{x}{5} - \frac{x}{2} < 3$

13. $-16x < -32$

Final Test page 5

14. $-\frac{2}{5}y < 20$

15. $5y - 7 > 4 - y$

16. You are told to create 50 cm^3 of a solution that is 8% bleach and water. How much pure bleach should you add to what amount of pure water?

17. You are instructed to fill a fish tank with 400 gallons of saltwater that is 25% salt. You are given two containers of different saltwater. One is 40% salt the other one is 20% salt. How much of each solution should you pour into the tank?

Final Test page 6

18. A 100 cm^3 solution of acid and water is 26% acid. How many cubic centimeters of pure water should be added to make a solution that is 13% acid?

19. A candy store wants to sell 4-pound bags of candied popcorn and peanuts at a price of $9.60 cents per pound. The candied popcorn costs 50 cents per ounce and the peanuts cost 65 cents per ounce. How many ounces of each ingredient should be mixed together to create such a mixture?
(Remember to keep your units the same).

Final Test page 7

20. A street vendor is selling balloons. He decides to create some bouquets of 15 balloons and sell them for $20 each. The regular latex balloons full of helium cost $1 each. The shiny Mylar balloons full of helium cost $1.50 each and the string, ribbons, and fancy weight cost $2 for each bouquet. How many of each balloon should he use?

21. Solve for the simultaneous solution by using the addition method.

$$x + 2y = -1$$
$$3x + 5y = -4$$

22. Solve for the simultaneous solution by using the substitution method.

$$x + 4y = 13$$
$$4x - y = 18$$

Final Test page 8

23. A plane flew from point A to point B against the wind for 4 hours. The plane returned back to point A in 3 hours. The distance between point A and point B is 1200 miles. What were the rate of the wind and the speed of the plane?

24. A train is traveling east at 80 mph. Another train is traveling west at 100 mph, just 90 miles away. How long will it take for the trains to meet and how far will each train have traveled?

Final Test page 9

25. A woman is trying to swim 2 miles upstream against the current and then back downstream in 30 minutes. If the rate of the current is 3 miles per hour, how fast should she swim?

26. In the last problem, you created a quadratic. Write it below.

____

Fill in the table below and then plot those points onto the graph.

$f(-2)$	$f(0)$	$f(2)$	$f(4)$	$f(5)$	$f(6)$	$f(8)$	$f(10)$

Final Test page 10

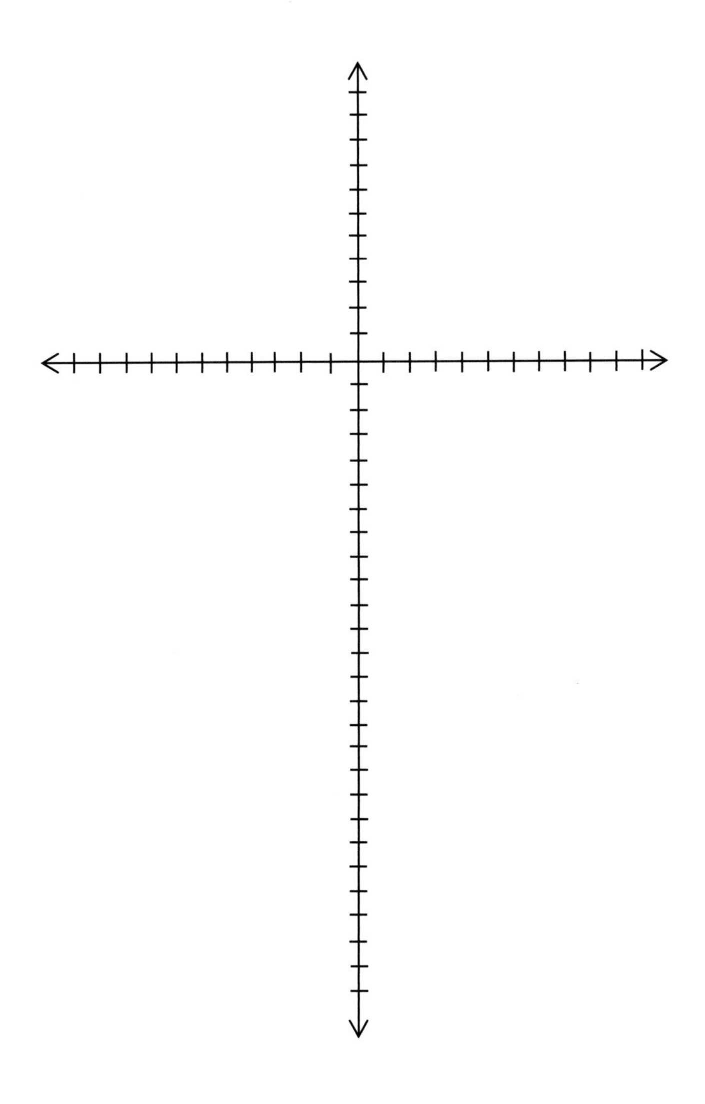

ANSWERS

ANSWERS: Worksheet 1

1. The auditorium can only hold 450 people. There are already 45 people working together to put on a play for their family and friends. Each person working on the play wants to sell tickets. How many tickets can each person sell without exceeding the total number of people allowed in the building at one time?

   What is the unknown? Number of tickets call it "x"
   Total number of people allowed? 450
   Number of people in the play? 45

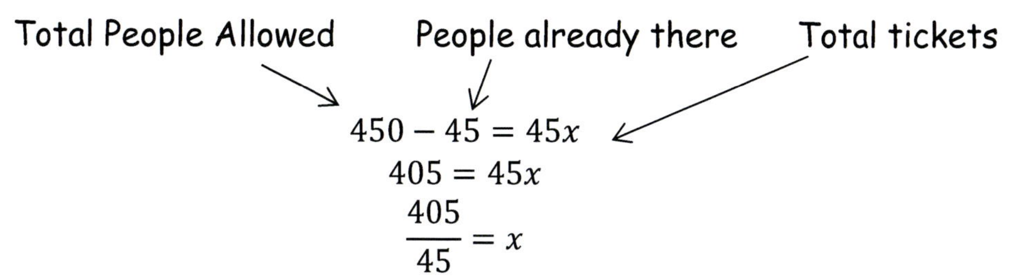

$$450 - 45 = 45x$$
$$405 = 45x$$
$$\frac{405}{45} = x$$

$9 = x$ Each member of the play can sell 9 tickets.

2. You are given 128 feet of chain link fence. It is your job to fence off a square shaped pen for several dogs. The pen must be 1024 square feet, in order for each dog to be comfortable. What should be the length of each side of the pen?

$$a = bh$$
$$1024 = b^2$$
$$\sqrt{1024} = \sqrt{b^2}$$
$$\sqrt{1024} = b \qquad \boldsymbol{b = 32}$$

Each side of the fence should be 32 feet

ANSWERS: Worksheet 1 page 2

3. An artist painted a big picture on the side of a building. The picture is in the shape of a circle with an area of 113.04 square meters. What is the diameter of the painted circle?

$$area\ of\ the\ circle = \pi r^2$$

$$113.04 = 3.14r^2$$

$$\frac{113.04}{3.14} = \frac{3.14r^2}{3.14}$$

$$36 = r^2$$

$$\sqrt{36} = \sqrt{r^2}$$

$$6 = r$$

The radius is 6 meters, so **the diameter is 12 meters.**

4. There is a lot of snow on the roads. A truck driver needs to put chains around his tires, so his truck doesn't slip in the snow. The tire measures 30 inches, from one side to the other. How long do the chains need to be, in order to go around the tires?

We are trying to find the circumference of the tire. The formula for finding the circumference of a circle is $c = d\pi$. The diameter is 30 inches, so let's do the math.

$$c = 30\pi$$

$$30\ inches \times 3.14 = 94.2\ inches$$

The chains need to be 94.2 inches, to fit around the tire.

5. There are 20 cubic feet of sand left over from a project. Mr. Wilson wants to build a sand box to hold the left-over sand. He built a rectangular box. The box is 5 feet long by 2 feet wide. How deep will the sand be once it is poured into the sandbox?

This is a volume question. The formula for volume is $v = bhw$. We already know the volume (20ft^3), the base (5 ft), and the width (2 ft). We need to solve for *h*. Let's do the math: $20 = 5 \times h \times 2$

$20 = 10h$ $\quad\quad$ $2 = h$ $\quad\quad$ **The sand is 2 feet deep.**

ANSWERS: Worksheet 2

1. It took 3 hours for the airplane to travel 1800 miles. What was the airplane's rate of speed?

$$\frac{d}{r} = t \qquad \frac{1800\,miles}{r} = 3\,hours$$

$$\frac{1800}{3} = r$$

$$\mathbf{600 = r} \qquad \boldsymbol{The\ airplane\ traveled\ 600\ miles\ per\ hour}$$

2. Teresa ran a 20-mile marathon. Her average speed was 5 miles per hour. Approximately how long did it take her to finish the race?

$$\frac{d}{r} = t \qquad \frac{20\,miles}{5\,mph} = t$$

$$\boldsymbol{4\ hours = t} \qquad \boldsymbol{It\ took\ Teresa\ 4\ hours\ to\ run\ the\ marathon}$$

3. John and Pat drove from Seattle to Portland. They drove an average of 50 miles per hour and it took about 3 hours to get there. How far did they travel?

$$\frac{d}{r} = t \qquad \frac{d}{50\,mph} = 3\,hours$$

$$d = 3 \times 50$$

$$\boldsymbol{d = 150\ miles} \qquad \boldsymbol{John\ and\ Pat\ traveled\ 150\ miles}$$

4. A golf ball travels 440 yards. It flew through the air at 100 miles per hour. How long did it take to land?

$$\frac{d}{r} = t \qquad \frac{440\,yards}{100\,mph} = t$$

$$1\,mile = 5280\,feet\ \ or\ 1760\,yards \qquad 440\,yards = .25\,mile$$

$$\frac{.25\,miles}{100\,mph} = t \qquad .0025\,hours$$

$$\boldsymbol{It\ took\ .0025\ hours\ for\ the\ ball\ to\ land.}$$

$$\boldsymbol{.0025 \times 60\ minutes = .15\ minutes} \qquad \boldsymbol{.15 \times 60\ seconds = 9\ seconds}$$

ANSWERS: Worksheet 3

1. Trinity walked to town at 4 mph. She rode the bus back at 40 mph. The town is 2 miles away. How long did it take Trinity to travel to town and back?

$$t = time$$

$$walked: \frac{2\ miles}{4\ mph} = t \qquad \frac{2}{4} = t \quad She\ walked\ for \frac{1}{2}\ hour$$

$$Bus\ ride: \frac{2}{40} = t \qquad \frac{1}{20} = t \qquad \frac{1}{20} \times \frac{60}{1} = \frac{6}{2}\ or\ 3\ minutes$$

Trinity's trip took 33 minutes.

2. A dog is resting against a pole to which he is attached with a 24 meter chain. He watches a cat approach to within 8 meters of the pole and promptly gives chase. The dog runs at 12 meters per second and the cat runs at 10 meters per second. Which does the dog reach first, the cat or the end of the chain?

To solve this problem, you will need to find how fast the cat can travel 16 meters and then how fast the dog can travel 24 meters. So we are searching for "time" in two different problems.

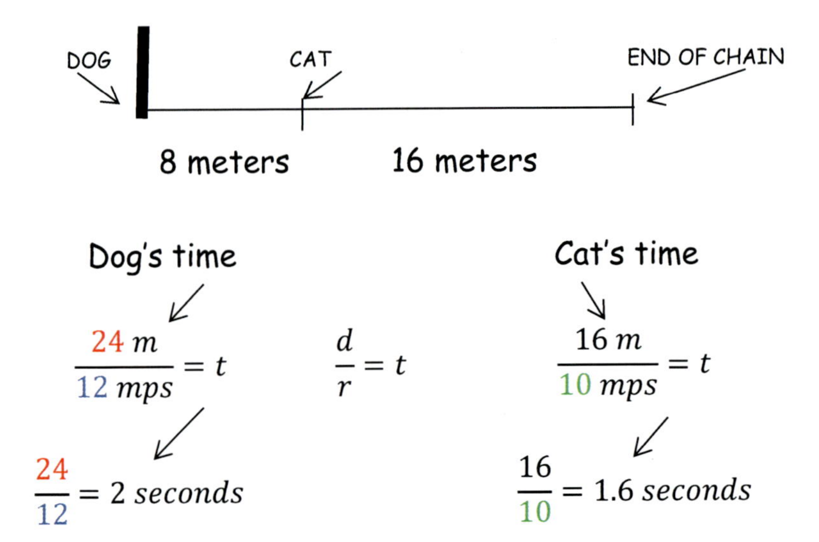

It took the dog 2 seconds to get to the end of the chain, but the cat was there in 1.6 seconds. The dog got to the end of the chain before he got the cat. Shew!

ANSWERS: Worksheet 3 page 2

3. A pilot flew north against the wind, which was blowing at 40 mph, for 5 hours. Returning south at the same speed, with the wind pushing him 40 mph, he made the return trip in 3 hours. What was the speed of the plane? (Hint: the distance is the same in both directions).

We are looking for the speed of the plane.

$$\frac{d}{r} = t$$

First, look at the 5-hour trip north. The plane was flying AGAINST the wind, so his speed (r) was slowed down by 40 mph. That is written as "r - 40" because whatever rate he was flying at, the wind slowed him down by 40 mph. We know the amount of time this trip took too; it was 5 hours.

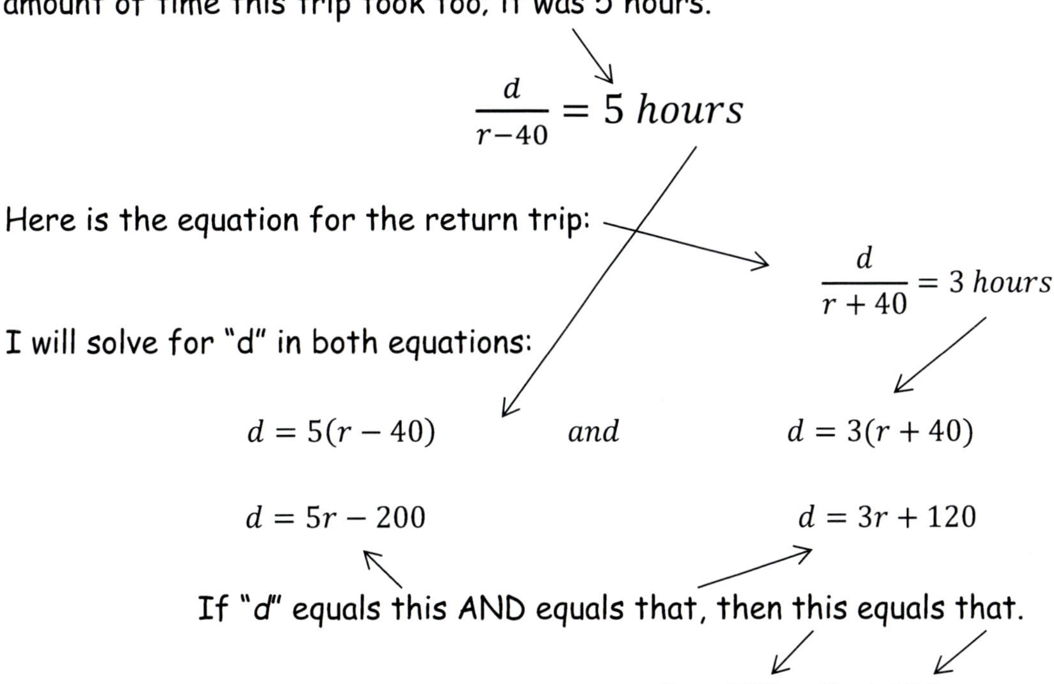

$$\frac{d}{r-40} = 5\ hours$$

Here is the equation for the return trip:

$$\frac{d}{r+40} = 3\ hours$$

I will solve for "d" in both equations:

$$d = 5(r-40) \qquad and \qquad d = 3(r+40)$$

$$d = 5r - 200 \qquad\qquad d = 3r + 120$$

If "d" equals this AND equals that, then this equals that.

$$5r - 200 = 3r + 120$$

Solve for "r" $\quad 5r - 3r - 200 = 120$

$$2r = 120 + 200$$

$$2r = 320 \qquad \boldsymbol{r = 160}$$

The plane was traveling at 160 mph.

ANSWERS: Worksheet 3 page 3

4. Two workers plan to mow 14 lawns. The older worker can mow 2 ½ lawns per hour. The younger worker can mow 1 lawn per hour. How long is this going to take the two workers to mow all 14 lawns.

$$\frac{14}{\frac{2.5}{1}+\frac{1}{1}}=x$$

$$\frac{14}{3.5}=x$$

$$4=x$$

It will take the two workers 4 hours to mow all 14 lawns.

ANSWERS: Worksheet 4

1. Below are two congruent triangles. Draw little lines to show which sides are congruent. Fill in the blanks with the appropriate values and then solve for x, y, and m.

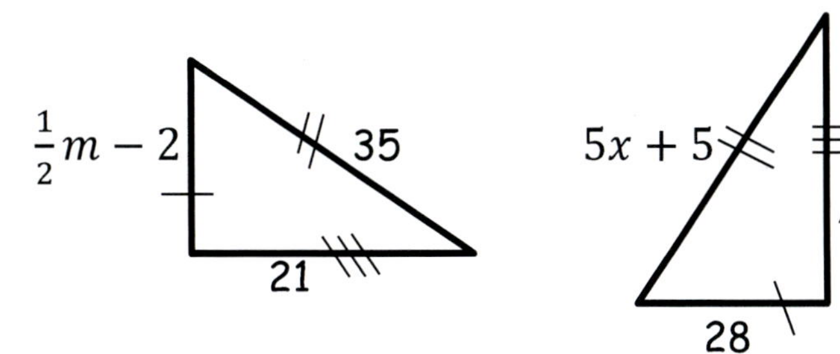

Side 1: $\mathbf{\frac{1}{2}m - 2 \cong 28}$

$\frac{1}{2}m \cong 28 + 2$

$\frac{1}{2}m \cong 30$ $\quad$ $\boldsymbol{m = 60}$

Side 2: $\mathbf{5x + 5 \cong 35}$

$5x \cong 35 - 5$

$5x \cong 30$ $\quad$ $\boldsymbol{x = 6}$

Side 3: $4y - 11 = 21$

$4y = 21 + 11$

$4y = 32$ $\quad$ $\boldsymbol{y = 8}$

2.

$$37 = 3x + (3^2 \cdot 2^2) - 4 \cdot \sqrt{64} + 12$$
$$37 = 3x + (36) - 4 \cdot \sqrt{64} + 12$$
$$37 = 3x + 36 - 4 \cdot 8 + 12$$
$$37 = 3x + 36 - 32 + 12$$
$$37 = 3x + 16$$
$$37 - 16 = 3x$$
$$21 = 3x$$

$$\mathbf{7 = x}$$

ANSWERS: Worksheet 4 page 2

3. Are the two triangles below congruent when $x = 4$?

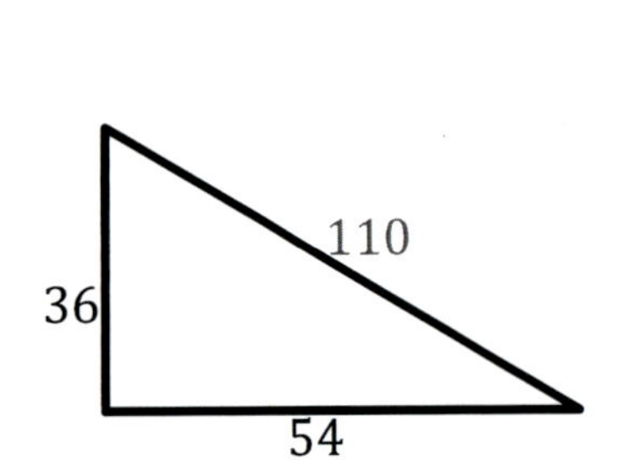

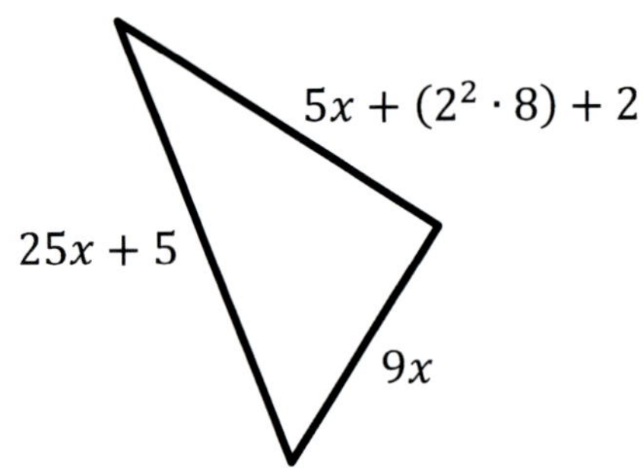

In order for the triangles to be congruent, the three corresponding sides must be equal. Replace x with 4 and see if all three sides are congruent.

$$\text{Does} \quad 5x + (2^2 \cdot 8) + 2 = 54 \quad \text{when } x = 4$$
$$20 + (32) + 2 = 54$$
$$54 = 54 \ \text{Yes, these sides are congruent}$$

$$\text{Does} \quad 25x + 5 = 110 \quad \text{when } x = 4$$
$$100 + 5 = 110$$
$$\textbf{No, it doesn't. These triangles are not congruent}$$

4. Every triangle has 3 angles. When you add all 3 angles together, it will always equal 180 degrees. Solve for x with the information given below.

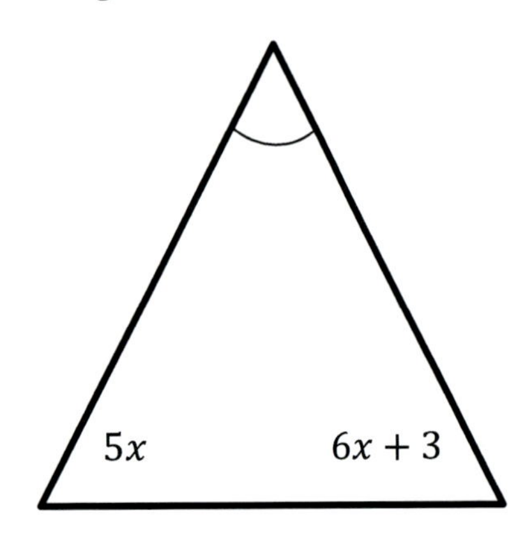

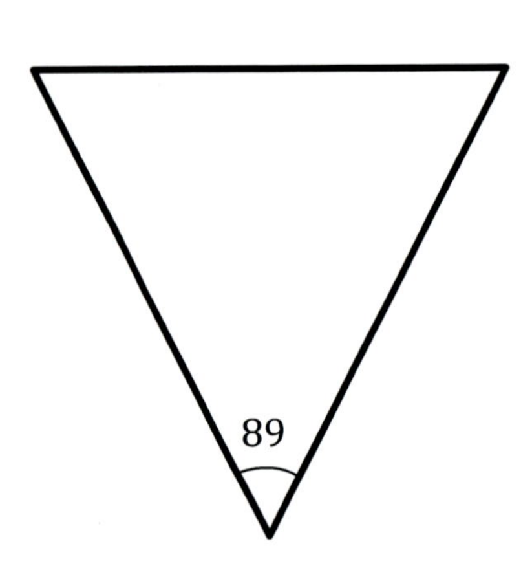

$$5x + (6x + 3) + 89 = 180$$
$$11x + 92 = 180$$
$$11x = 88 \qquad \boldsymbol{x = 8}$$

ANSWERS: Worksheet 5

1.

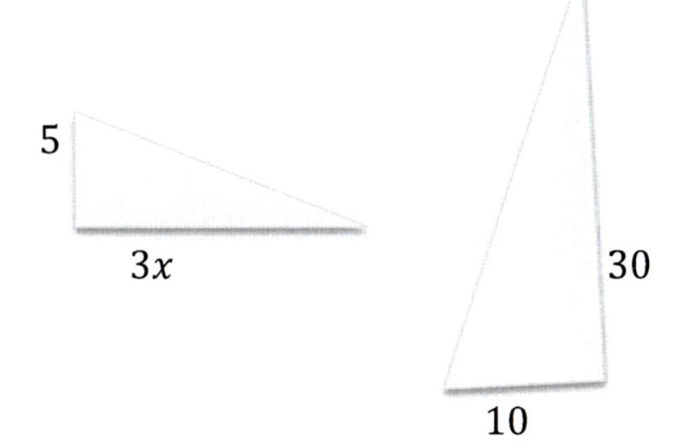

$$\frac{5}{3x} = \frac{10}{30}$$

$$30x = 150$$

$$x = \frac{150}{30}$$

$$\boldsymbol{x = 5}$$

2.

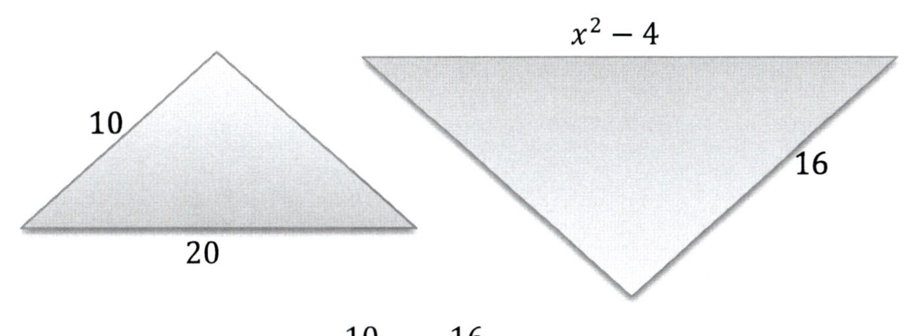

$$\frac{10}{20} = \frac{16}{x^2 - 4}$$

$$320 = 10x^2 - 40$$

$$360 = 10x^2$$

$$36 = x^2 \qquad \boldsymbol{6 = x}$$

3.

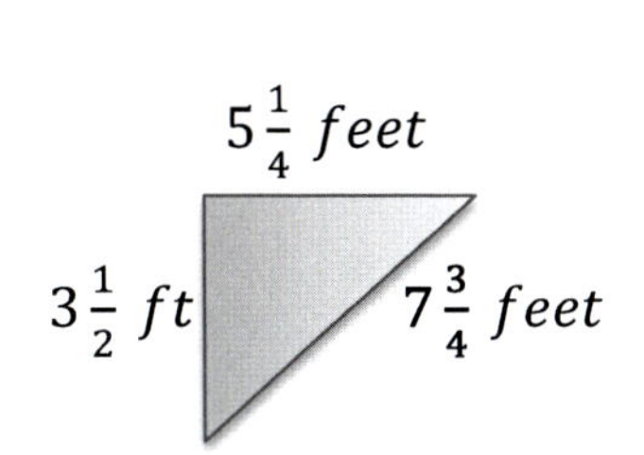

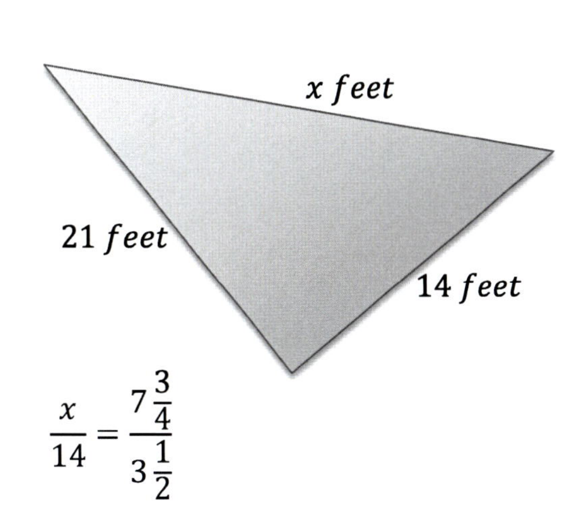

$$\frac{x}{14} = \frac{7\frac{3}{4}}{3\frac{1}{2}}$$

ANSWERS: Worksheet 5 page 2

$$3\frac{1}{2}x = \frac{14}{1}\cdot\frac{31}{4}$$

$$3\frac{1}{2}x = \frac{434}{4}$$

$$x = \frac{\frac{434}{4}}{3\frac{1}{2}} \leftarrow \boxed{\frac{434}{4}\cdot\frac{2}{7} = 31}$$

$$\boldsymbol{x = 31\ feet}$$

4. Solve for x in these two similar, isosceles Triangles.

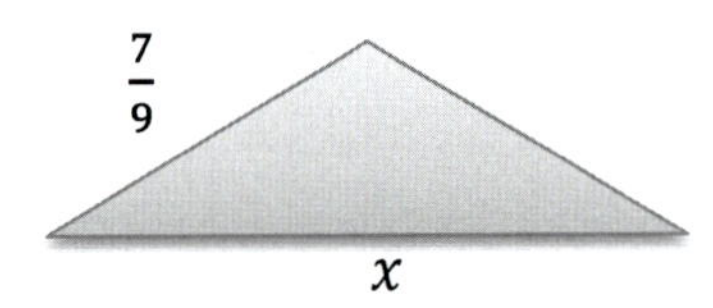

An isosceles triangle has two equal sides.

$$\frac{\frac{7}{9}}{x} = \frac{\frac{18}{7}}{6}$$

$$\frac{18}{7}x = \frac{6}{1}\cdot\frac{7}{9}$$

$$\frac{18}{7}x = \frac{42}{9}$$

$$x = \frac{\frac{42}{9}}{\frac{18}{7}}$$

$$x = \frac{42}{9}\cdot\frac{7}{18}$$

$$x = \frac{294}{162} \qquad \boldsymbol{x = \frac{49}{27}}$$

ANSWERS: Worksheet 5 page 3

5. A builder wants to know the square footage of the shaded triangular area of the building below. His ladder isn't tall enough to measure the height of the triangle, so he will have to use math instead. He knows that the pitch of the roof is 5:12. Use similar triangles to find the area of the shaded area below.

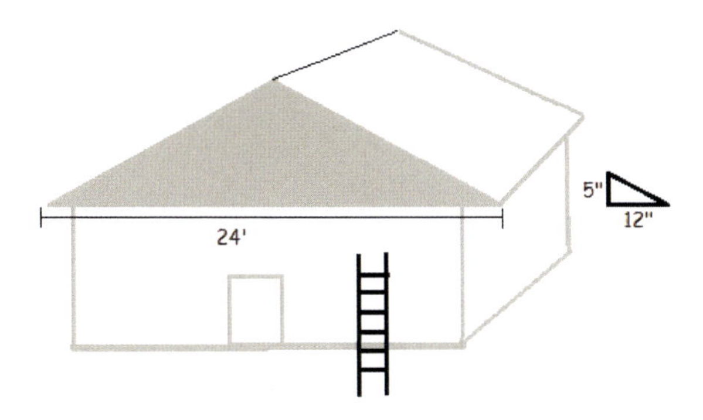

This problem requires some extra thinking. We are looking for the area of the shaded triangle. In order to find the area, we will need to know the height. The pitch of the roof is a similar triangle. It shows that the roof will go up 5" for every 12 horizontal inches. We will use that ratio to create a proportion, so we can find the height of the shaded triangle.

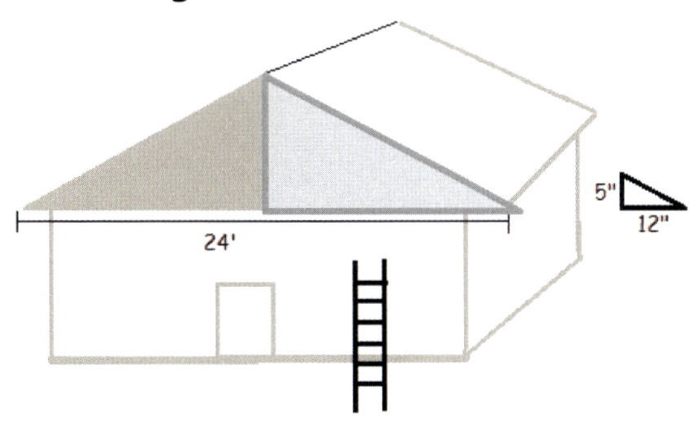

The base of the shaded triangle is 12 feet. Our little triangle is 12 inches OR feet. Let's go with feet! Here are the ratios. Solve for x in the proportion.

$$\frac{5}{12} = \frac{x}{12}$$

$$12x = 60$$

$$x = \frac{60}{12} \qquad x = 5 \text{ feet}$$

ANSWERS: Worksheet 5 page 4

The height of the shaded triangle is 5 feet.

Now we can solve for the area of the shaded triangle by filling in the Area Formula.

$$A = \frac{1}{2}bh \qquad A = \frac{1}{2} \cdot 24\, feet \cdot 5\, feet$$

$$A = 60\, ft^2$$

The area of the triangle is 60 square feet.

ANSWERS: Chapter 1 Test

1. Judy needs to receive an average score of 94 in order to earn a scholarship. On the first test she received a score of 95. On the second test, she earned a score of 93. The third test was her worst score, it was a 92. What score must she get on her fourth test in order to earn the scholarship?

$$95 + 93 + 92 + x = 94 \cdot 4$$
$$280 + x = 376$$
$$x = 376 - 280$$
$$\boldsymbol{x = 96}$$

Judy needs to get a score of 96.

2. Daryl wants to save up $2000 to get his truck painted. He already has $750 saved up and he can earn $40.00 per hour at his job. How many hours does he need to work to earn enough money for the new paint job?

$$2000 = 750 + 40x$$
$$2000 - 750 = 40x$$
$$1250 = 40x$$
$$\boldsymbol{31.25 = x}$$

Daryl needs to work 31.25 hours.

3. A horse is tethered to a pole in the center of a pen. The horse walks in a circle, stretching the rope as far as possible as he walks around the pole. Each time he circles the pole he travels 113.04 feet. How long is the rope?

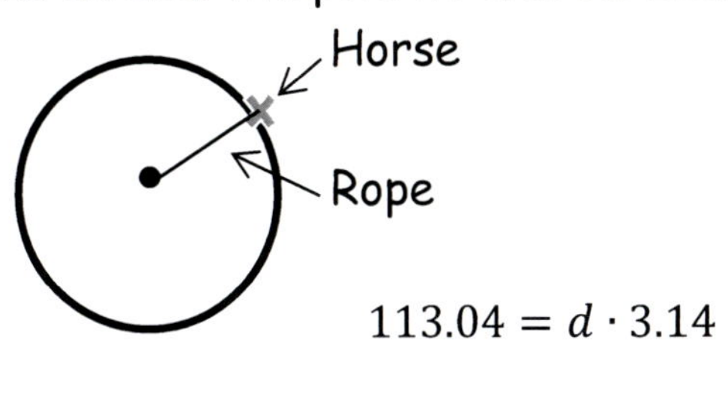

$$113.04 = d \cdot 3.14$$

$$\frac{113.04}{3.14} = d$$

$$36 = d$$

If the diameter is 36 feet, then the rope is 18 feet.

ANSWERS: Chapter 1 Test page 2

4. A homemade rocket shot straight up into the air. The radar detector showed that it traveled at 110 mph. The stop watch showed that it traveled upwards for 1 minute before it started to fall. Approximately how far did the rocket travel?

$$\frac{d}{110} = \frac{1}{60}$$

$$110 = 60d$$

$$\frac{110}{60} = d$$

$$\frac{11}{6} = distance$$

The rocket traveled $1\frac{5}{6}$ of a mile.

5. A train is traveling east at 100 mph. Another train is heading straight towards it at 80 mph, 55 miles away. The trains will hit each other at the halfway point, if the tracks are not switched in time. The engineer can't switch the tracks for 15 minutes. Will the trains hit each other before the engineer can flip the switch?

First Train

$$\frac{x}{100} = \frac{1}{4}$$

$$100 = 4x$$

$$x = 25$$

Second Train

$$\frac{x}{80} = \frac{1}{4}$$

$$80 = 4x$$

$$20 = x$$

Halfway

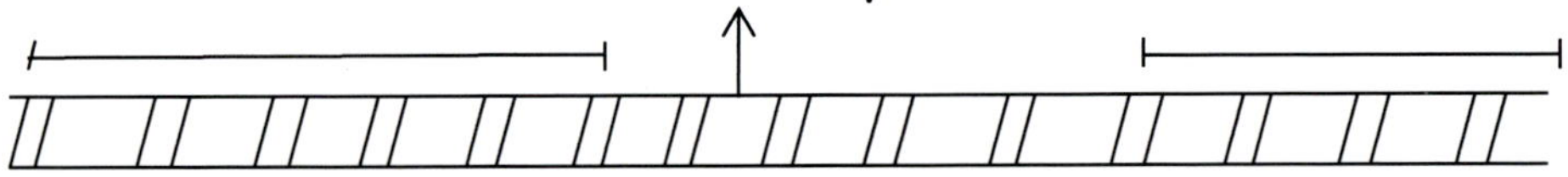

ANSWERS: Chapter 1 Test page 3

In 15 minutes $\left(\frac{1}{4}\ hour\right)$ the first train will have traveled 25 miles and the second train will have traveled 20 miles. The engineer will have switched the tracks before the trains hit.

6. Are the two isosceles triangles below congruent?

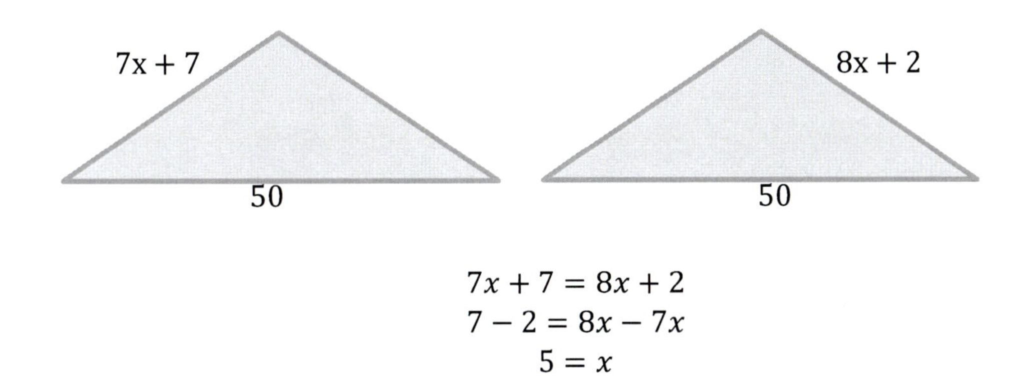

$$7x + 7 = 8x + 2$$
$$7 - 2 = 8x - 7x$$
$$5 = x$$

If x = 5, then both sides equal 42. Since these are isosceles triangles, the two sides are equal, so these are congruent triangles.

7. A builder needs to replace all of the blue boards on the house below. He charges $10 per foot, to replace the boards. How much will the builder charge?

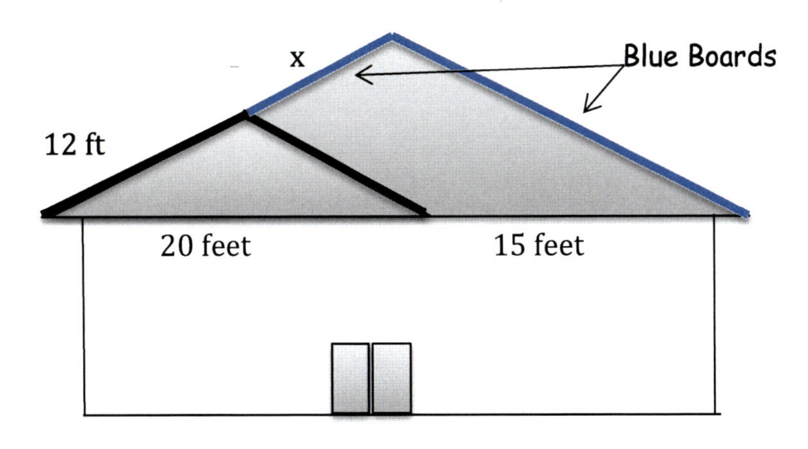

ANSWERS: Chapter 1 Test page 4

$$\frac{12}{20} = \frac{12 + x}{35}$$

$$420 = 240 + 20x$$

$$420 - 240 = 20x$$

$$180 = 20x$$

$$9 = x$$

The shorter blue board is 9 feet long and the longer blue board is 21 feet long. The builder charges $10 per foot to replace the boards.

$$30 \cdot \$10 = \$300$$

The builder will charge $300 to replace the boards.

8. A factory can produce 2 guitars every 5 hours during the day shift and 4 guitars in 5 hours during the night shift. If each shift works 12 hours per day, how many days will it take to fill an order of 72 guitars?

$$\frac{72}{\frac{2}{5} + \frac{4}{5}} = x$$

$$\frac{72}{\frac{6}{5}} = x$$

$$\frac{72}{1} \times \frac{5}{6} = x$$

$$\frac{360}{6} = x$$

$$60 = x$$

ANSWERS: Chapter 1 Test page 5

It will take 60 hours for both shifts to complete the order. Each shift will work 12 hours per day, so divide 60 by 12 to find the number of days.

$$60 \div 12 = 5\ days$$

It will take 5 days to complete the order.

9. Jo-Jo is going to drive from Tacoma to Federal Way, drop off an envelope and then drive back to Tacoma. This trip will take 40 minutes to complete. If he continues at the same rate of speed, how long will it take him to drive to Kent?

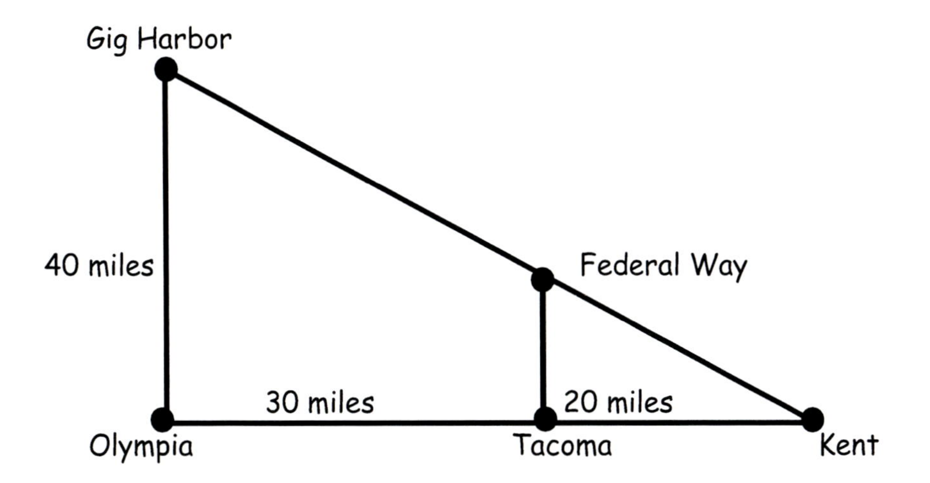

The distance between Federal Way and Tacoma is unknown. The two triangles are similar, so let's use proportions to find that distance.

$$\frac{40}{50} = \frac{x}{20}$$

$$50x = 800$$

$$x = 16$$

ANSWERS: Chapter 1 Test page 6

The distance between Federal Way and Tacoma is 16 miles. Jo-Jo drove there and back for a total of 32 miles. It took him 40 minutes, so let's find out his average rate of speed.

$$\frac{32}{x} = 40\ minutes$$

I can write "40 minutes" a couple different ways. I can call it $\frac{40}{60}$ or I can call it $\frac{2}{3}$ of an hour. I prefer the reduced fraction.

$$\frac{32}{x} = 40\ minutes$$

$$\frac{32}{x} = \frac{2}{3}$$

$$2x = 96 \qquad x = 48$$

Jo-Jo drove an average of 48 miles per hour. Kent is 20 miles away. Let's figure out how long it will take him to get to Kent at that speed.

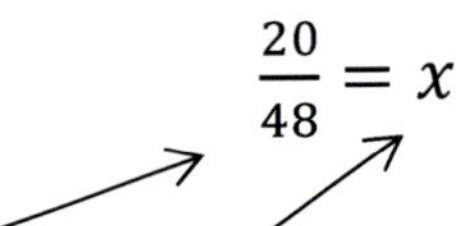

Since this is "miles per hour" and this is going to be "minutes," I can solve it two different ways. I can say $\frac{20}{48} = \frac{x}{60}$ or I can just reduce the fraction, knowing that the answer will be a fraction of an hour.

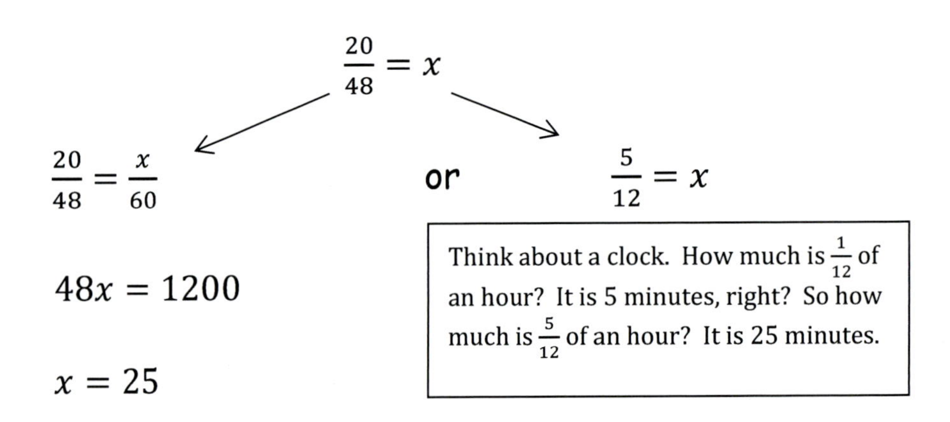

It will take Jo-Jo 25 minutes to drive to Kent.

ANSWERS: Worksheet 6

1. In a box of 100 animal cookies, there are 4 different animal shapes. There are 25 bears, 20 giraffes, 30 lions, and 25 camels. What is the probability of randomly selecting a lion shaped cookie out of the box?

$$\frac{30}{100} = \mathbf{\frac{3}{10}}$$

2. I have written down some names and put them in a hat. Six of them are girl names and seven of them are boy names. If I draw a name out of the hat, what is the probability of it being a boy's name?

$$\mathbf{\frac{7}{13}}$$

3. There are a dozen light bulbs in a box. Three of them are burnt out. What is the probability of selecting two light bulbs that both work?

$$\frac{9}{12} \cdot \frac{8}{11} = \frac{72}{132} = \mathbf{\frac{6}{11}}$$

4. I roll two dice at the same time. What is the probability of them both landing on 5?

$$\frac{1}{6} \cdot \frac{1}{6} = \mathbf{\frac{1}{36}}$$

5. Two cards are selected from an ordinary deck of cards. What is the probability of them both being black or both being red?

$$\frac{1}{2} \cdot \frac{25}{51} = \frac{25}{102} \qquad \frac{1}{2} \cdot \frac{25}{51} = \frac{25}{102}$$

Probability of them both being black

Probability of them both being red

$$\frac{25}{102} + \frac{25}{102} = \frac{50}{102} = \mathbf{\frac{25}{51}}$$

Black OR Red

ANSWERS: Worksheet 7

1. $x - 8 > 5$ → $\boldsymbol{x > 13}$

2. $-10x < 20$ → $\frac{-10x}{-10} < \frac{20}{-10}$

   $\boldsymbol{x > -2}$

3. $-2x > -12$ → $\frac{-2x}{-2} > \frac{-12}{-2}$

   $\boldsymbol{x < 6}$

4. $5x - 4 < 2x + 8$ → $5x - 2x < 8 + 4$

   $3x < 12$

   $\boldsymbol{x < 4}$

5. $\frac{3x-1}{2} < 10$

   $\frac{\cancel{2}}{1} \cdot \frac{3x-1}{\cancel{2}} < 10 \cdot 2$

   $3x - 1 < 20$

   $3x < 21$

   $\boldsymbol{x < 7}$

6. $4x - 3 < 2x + 7$

   $4x - 2x < 7 + 3$

   $2x < 10$

   $\boldsymbol{x < 5}$

7. $\frac{3x+5}{2} < 7$

   $\frac{\cancel{2}}{1} \cdot \frac{3x+5}{\cancel{2}} < 7 \cdot 2$

   $3x + 5 < 14$

   $3x < 14 - 5$

   $3x < 9$

   $\boldsymbol{x < 3}$

ANSWERS: Worksheet 8

1. A container holds 18 liters of a solution that is 7% acid. It is poured into 12 liters of a 20% acid and water solution. What is the percentage of acid in the new solution?

18 liters + 12 liters = 30

$$(18 \times .07) + (12 \times .2) = 30x$$

$$1.26 + 2.4 = 30x$$

$$3.66 = 30x$$

$$\frac{3.66}{30} = x \qquad x = 0.122$$

$$\boldsymbol{x = 12.2\%}$$

2. You have a container with 32 ounces of a solution of 23% alcohol and water. Your assistant accidentally pours 8 ounces of an alcohol and water solution and now your container is down to 20% alcohol. What was the percentage of alcohol in your assistant's glass?

$$(32 \times .23) + (8 \times x) = 40 \times .2$$

$$7.36 + 8x = 8$$

$$8x = 8 - 7.36$$

$$8x = .64$$

$$x = .08 \qquad \boldsymbol{x = 8\%}$$

3. How much of a 5% iodine and water solution should be added to 4 cups of 2% iodine and water to create a solution that is 3% iodine and water?

$$(.05 \times x) + (4 \times .02) = .03(4 + x)$$

$$.05x + .08 = .12 + .03x$$

$$.05x - .03x = .12 - .08$$

$$.02x = .04$$

$$\boldsymbol{x = 2\ cups}$$

ANSWERS: Worksheet 8 page 2

4. How much pure water should be added to 180 ml solution of bleach and water that is 23% bleach to dilute it down to a solution that is 10% bleach and water?

$$(0 \times x) + (180 \times .23) = .10(180 + x)$$

$$0 + 41.4 = 18 + .1x$$

$$41.4 - 18 = .1x$$

$$23.4 = .1x$$

$$\mathbf{234\ ml = x}$$

5. A gas can contains twelve cups of an oil and gasoline solution. It is currently 60% oil, but it needs to be down to 45%. How much pure gas should be added to the solution to get it to 45% oil?

$$.6(12) + (0 \times x) = .45(12 + x)$$

$$7.2 + 0 = 5.4 + .45x$$

$$7.2 - 5.4 = .45x$$

$$1.8 = .45x$$

$$\mathbf{x = 4\ cups}$$

6. If you mix 135 cm^3 of a solution that is 17% alcohol and water with 80 cm^3 of a 53% alcohol and water solution, what will the new percentage of alcohol?

$$.17(135) + .52(80) = 215x$$

$$22.95 + 41.6 = 215x$$

$$64.55 = 215x$$

$$x = .30 \qquad \mathbf{x = 30\%\ alcohol}$$

ANSWERS: Worksheet 9

1. In this part of the country, the water in a car's radiator should be 20% anti-freeze, so it doesn't freeze during the winter months. The radiator holds 5 gallons of water and anti-freeze solution. How much pure water and pure anti-freeze should be used to obtain the proper solution in the car?

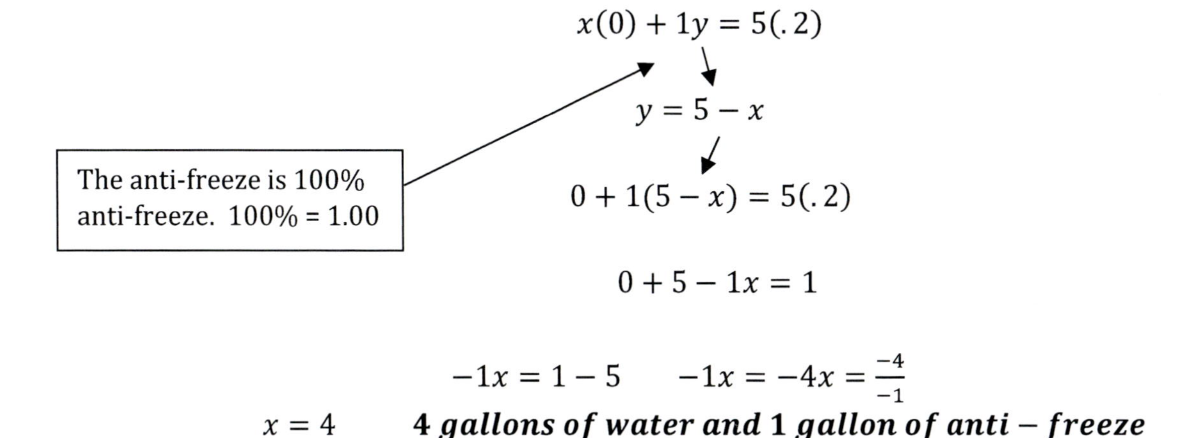

$$x(0) + 1y = 5(.2)$$

$$y = 5 - x$$

$$0 + 1(5 - x) = 5(.2)$$

$$0 + 5 - 1x = 1$$

$$-1x = 1 - 5 \qquad -1x = -4x = \frac{-4}{-1}$$

$x = 4$ $\qquad$ $\boldsymbol{4\ gallons\ of\ water\ and\ 1\ gallon\ of\ anti-freeze}$

2. A nurse has two containers of Isopropyl Rubbing Alcohol. One container is 70% alcohol and the other is 91%. She is instructed to use just enough of each solution to obtain 50 milliliters of 80% Isopropyl Rubbing Alcohol. How much of each container should she use? Round your answers to the nearest one-hundredth.

$$x(.70) + y(.91) = 50(.80)$$

$$.7x + .91(50 - x) = 40$$

$$.7x + .91(50 - x) = 40$$

$$.7x + 45.5 - .91x = 40$$

$$.7x - .91x = 40 - 45.5$$

$$-.21x = -5.5 \qquad x = \frac{-5.5}{-.21} \qquad x = 26.19$$

She should use 26.19 milliliters of the 70% and 23.81 milliliters of the 91% alcohol.

ANSWERS: Worksheet 9 page 2

3. A nurse is preparing an IV bag of water and sugar. She has a large bottle that is 15% sugar, but the IV bag needs to be one liter of 9% sugar water. How much of the bottle should she add to what amount of water?

$$x(.15) + y(0) = 1(.09)$$

$$.15x + 0 = .09$$

$$.15x = .09$$

$$x = \frac{.09}{.15} \qquad x = .6$$

She should add .6 liters of the 15% sugar solution to .4 liters of pure water.

4. A one-gallon tub of Epsom Salt and water is 6% Epsom salt. How much pure water should be added to make a solution that is 4% Epsom salt?

$$1 - gallon(.06) + x(0) = .04(y)$$

$$128\ ounces(.06) + x(0) = .04(128 + x)$$

$$7.68 + 0 = 5.12 + .04x$$

$$7.68 - 5.12 = .04x$$

$$2.56 = .04x$$

$$\frac{2.56}{.04} = x \qquad x = 64\ ounces$$

She should add 64 ounces of pure water.

ANSWERS: Worksheet 9 page 3

5. A solution of acid and water is 16% acid. How many cups of this solution should be mixed with pure water to make 21 cups of a solution that is 10% acid?

$$.16(x) + 0(y) = 21(.1)$$

$$x + y = 21$$

$$x = 21 - y$$

$$.16(21 - y) + 0y = 21(.1)$$

$$3.36 - .16y + 0 = 2.1$$

$$-.16y + 0 = 2.1 - 3.36$$

$$-.16y = -1.26$$

$$y = \frac{-1.26}{-.16} \qquad y = 7.875$$

7.875 cups of pure water should be added.

6. A masonry needs to clean a concrete wall. A solution of bleach and water works very well. He will need 5 gallons of a bleach water solution that is 20% bleach. He has a container of 12% bleach water. How much of that solution should he mix with pure bleach to get the proper percentage?

$$.12(x) + 1.00(y) = 5(.20)$$
$$x + y = 5$$
$$y = 5 - x$$

$$.12x + 1(5 - x) = 5(.2)$$

$$.12x + 5 - 1x = 1.0$$
$$.12x - 1x = 1 - 5$$

$$-.88x = -4 \qquad x = \frac{-4}{-.88} \qquad x = 4.5454$$

$$\boldsymbol{x = 4.5454\ gallons\ of\ 12\%\ solution}$$

ANSWERS: Worksheet 10

1. We have two types of Christmas cards. One set has a glitter border; they sell for 50 cents each. The other set is plain and only cost 30 cents each. How many of each should be placed in assortment boxes of 50 cards each to sell for $18 per box?

$$.3x + .5(50 - x) = 18$$

$$.3x + 25 - .5x = 18$$

$$-.2x + 25 = 18$$

$$-.2x = 18 - 25$$

$$-.2x = -7$$

$$x = \frac{-7}{-.2}$$

$$-7 \div -.2 = 35$$

$$\boldsymbol{x = 35}$$

The box of 50 cards should contain 35 of the 30 cent cards and 15 of the 50 cent cards.

2. A dairy sells pure cream for 75 cents per cup and skim milk for 25 cents per cup. How many cups of each should be mixed to make coffee creamer that will sell for $8 per gallon?

There are 16 cups in a gallon, so we know that

$$x + y = 16 \quad and \quad y = 16 - x$$

$$.75x + .25(16 - x) = \$8$$

$$.75x + 4 - .25x = 8$$

ANSWERS: Worksheet 10 page 2

$$.5x = 8 - 4$$

$$.5x = 4$$

$$x = 8$$

$$y = 16 - x \qquad so\ y = 8$$

The dairy should mix 8 cups of each ingredient.

3. A store owner is selling fresh cut fruit in $4\frac{1}{2}$ pound containers. The berries cost $4.75 per pound and the melon costs $4.20 per pound. The store owner plans to add $5 per container for the labor, materials, and profit. If he wants to sell the containers of fruit for $25 each, how many pounds of each fruit should he put in each mixture?

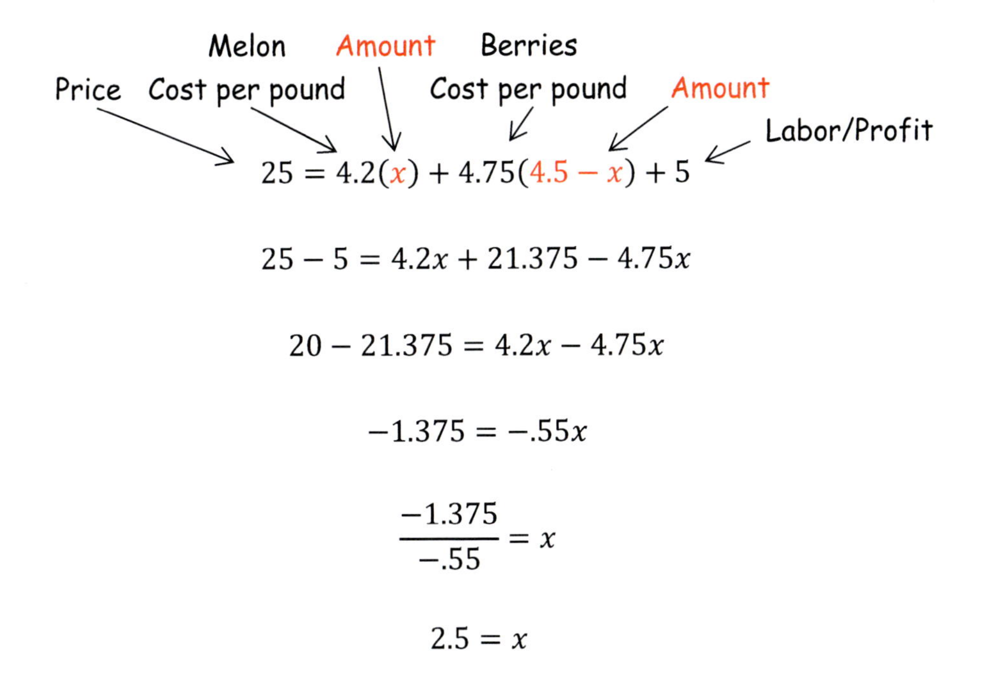

$$25 = 4.2(x) + 4.75(4.5 - x) + 5$$

$$25 - 5 = 4.2x + 21.375 - 4.75x$$

$$20 - 21.375 = 4.2x - 4.75x$$

$$-1.375 = -.55x$$

$$\frac{-1.375}{-.55} = x$$

$$2.5 = x$$

The store owner should use 2.5 pounds of melon and 2 pounds of berries.

ANSWERS: Chapter 2 Test

1. A number is randomly selected from the set of numbers below.

$$5, 11, 23, 42$$

What is the probability of that number being greater than 25?

$$\frac{\mathbf{1}}{\mathbf{4}}$$

2. A drawer holds 12 socks. There are 6 blue socks and 6 black socks. What is the probability of selecting two socks of the same color?

This problem can be solved two different ways. First, I'll solve it as if I'm reaching in the drawer and selecting any color sock and then grabbing another sock in hopes of getting a match.

$$\frac{12}{12} \cdot \frac{5}{11} = \frac{60}{132} = \frac{\mathbf{5}}{\mathbf{11}}$$

Next, I will solve it by getting the probability of grabbing two blue socks and then I'll ADD that to the probability of selecting two black socks.

$$\frac{6}{12} \cdot \frac{5}{11} = \frac{30}{132} = \frac{5}{22} \leftarrow \text{Probability of 2 blue socks}$$

$$\frac{6}{12} \cdot \frac{5}{11} = \frac{30}{132} = \frac{5}{22} \leftarrow \text{Probability of 2 black socks}$$

$$\frac{5}{22} + \frac{5}{22} = \frac{10}{22} = \frac{\mathbf{5}}{\mathbf{11}}$$

Solve the inequalities.

3. $4x - 6 < 7x + 12$

$$4x - 7x < 12 + 6$$
$$-3x < 18$$
$$\boldsymbol{x > -6}$$

ANSWERS: Chapter 2 Test page 2

4. $-6 < -2x + 4$

$$-6 - 4 < -2x$$

$$-10 < -2x$$

$$\frac{-10}{-2} > x$$

$$\mathbf{5 > x}$$

5. $\frac{4x+6}{2} < 4$

$$4x + 6 < 4 \cdot 2$$

$$4x < 8 - 6$$

$$4x < 2$$

$$x < \frac{2}{4}$$

$$\boldsymbol{x < \frac{1}{2}}$$

6. A solution of acid and water is 16% acid. How many cups of this solution should be mixed with pure water to make 32 cups of a solution that is 10% acid?

$$.16x + 0y = 32(.1)$$

$$\boldsymbol{x + y = 32}$$
$$\boldsymbol{x = 32 - y}$$
$$.16(32 - y) = 32(.1)$$
$$5.12 - .16y = 3.2$$
$$-.16y = 3.2 - 5.12$$
$$-.16y = -1.92$$
$$y = \frac{-1.92}{-.16} \qquad y = 12 \qquad x = 20$$

You should mix 20 cups of the solution with 12 cups of pure water

ANSWERS: Chapter 2 Test page 3

7. A solution of water and lemon juice will be mixed to make 8 cups of lemonade. We don't want the drink to be too sweet or too sour, so it must be 15% lemon flavoring. How much pure water and pure lemon juice should we use?

$$1(x) + 0(Y) = 8(.15)$$

$$x + y = 8$$

$$x = 8 - y$$

$$1(8 - y) + 0(Y) = 8(.15)$$

$$8 - 1y + 0 = 1.20$$

$$-y = 1.2 - 8$$

$$-y = -6.8$$

6.8 cups of pure water and 1.2 cups of lemon juice.

8. A solution of salt and water is 30% salt. How many cm^3 of this solution should be mixed with pure water to make 18 cm^3 of a solution that is 10% salt?

$$.3x + y(0) = 18(.1)$$

$$x + y = 18$$
$$x = 18 - y$$

$$.3(18 - y) + y(0) = 18(.1)$$

$$5.4 - .3y + 0 = 1.8$$

$$-.3y = -3.6$$

$$y = \frac{-3.6}{-.3} \qquad y = 12 \qquad x = 6$$

ANSWERS: Chapter 2 Test page 4

9. The owner of a pizza store wants to write down the exact recipe for a pepperoni and olive pizza, so he is sure to keep the cost and weight the same for every pizza. He already knows how much a cheese pizza costs and weighs and would like to add 8 ounces of toppings for only $1.80 more per pizza. The pepperoni costs 90 cents per ounce and the olives cost 10 cents per ounce. How many ounces of each ingredient should the store owner use?

$$.9x + .1y = \$1.80$$

$$x + y = 8.0$$

$$y = 8 - x$$

$$.9x + .1(8 - x) = 1.8$$

$$.9x + .8 - .1x = 1.8$$

$$.8x = 1.8 - .8$$

$$.8x = 1$$

$$x = \frac{1}{.8}$$

$$x = \frac{1}{1} \div \frac{8}{10} = \frac{10}{8} \; or \; \boldsymbol{1\frac{1}{4}} \; \boldsymbol{ounces\ of\ pepperoni}$$

$$8 - 1\frac{1}{4} = \boldsymbol{6\frac{3}{4}} \; \boldsymbol{ounces\ of\ olives}$$

ANSWERS: Chapter 2 Test page 5

10. Debbie wants to make snack mix and then fill a 4 cup bag that cost $5.50 per bag. The cereal costs $0.50 per cup and the nuts cost $3.00 per cup. The bag, butter, and seasoning costs 1.00 per 4 cup bag. How much cereal and nuts should be added to each bag to keep the cost at $5.50?

$$x + y = 4\ cups$$

$$y = 4 - x$$

$$.5x + 3y + 1 = 5.50$$

$$.5x + 3(4 - x) + 1 = 5.5$$

$$.5x + 12 - 3x + 1 = 5.5$$

$$.5x - 3x = 5.5 - 13$$

$$-2.5x = -7.5$$

$$x = 3$$

Debbie should use 3 cups of cereal and 1 cup of nuts

ANSWERS: Worksheet 11

1. By Addition:

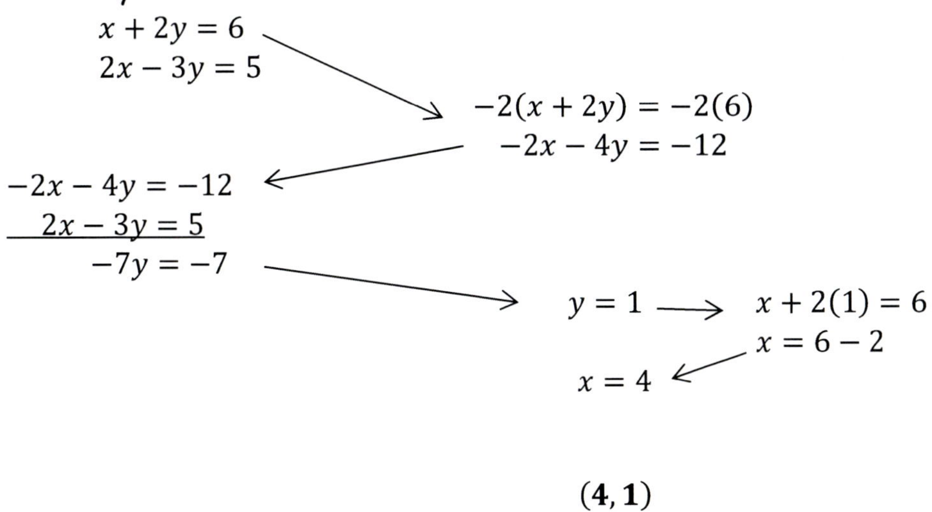

$\mathbf{(4, 1)}$

Substitution Method:

$x + 2y = 6$

$2x - 3y = 5$

$x = 6 - 2y$

$\mathbf{2(6 - 2y) - 3y = 5}$

$\mathbf{12 - 4y - 3y = 5}$

$\mathbf{-7y = 5 - 12}$

$\mathbf{-7y = -7}$

$\mathbf{y = 1}$

$\mathbf{x + 2(1) = 6}$

$\mathbf{x = 6 - 2}$

$\mathbf{x = 4}$ $\quad$ $\mathbf{(4, 1)}$

ANSWERS: Worksheet 11 page 2

2. By addition:

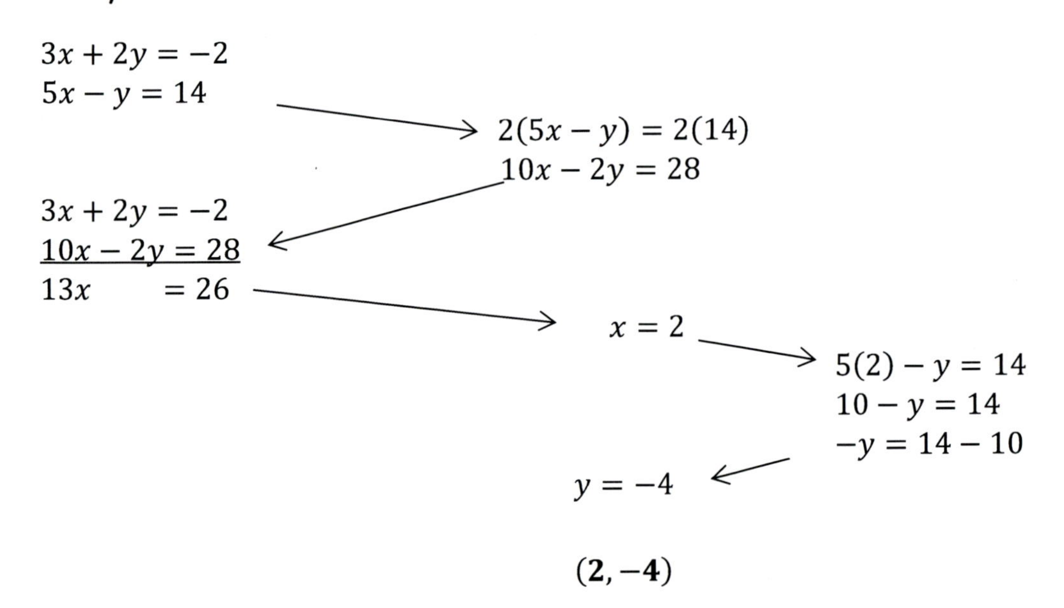

By Substitution:

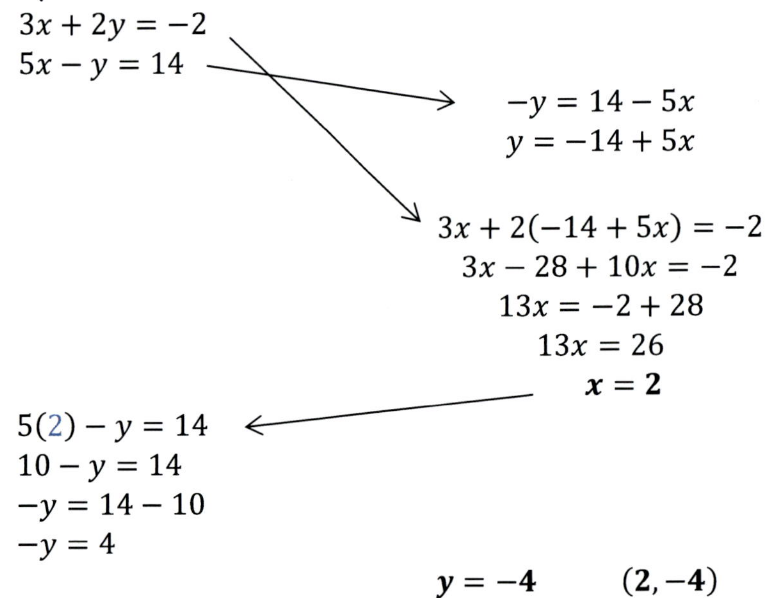

ANSWERS: Worksheet 11 page 3

3. By addition:

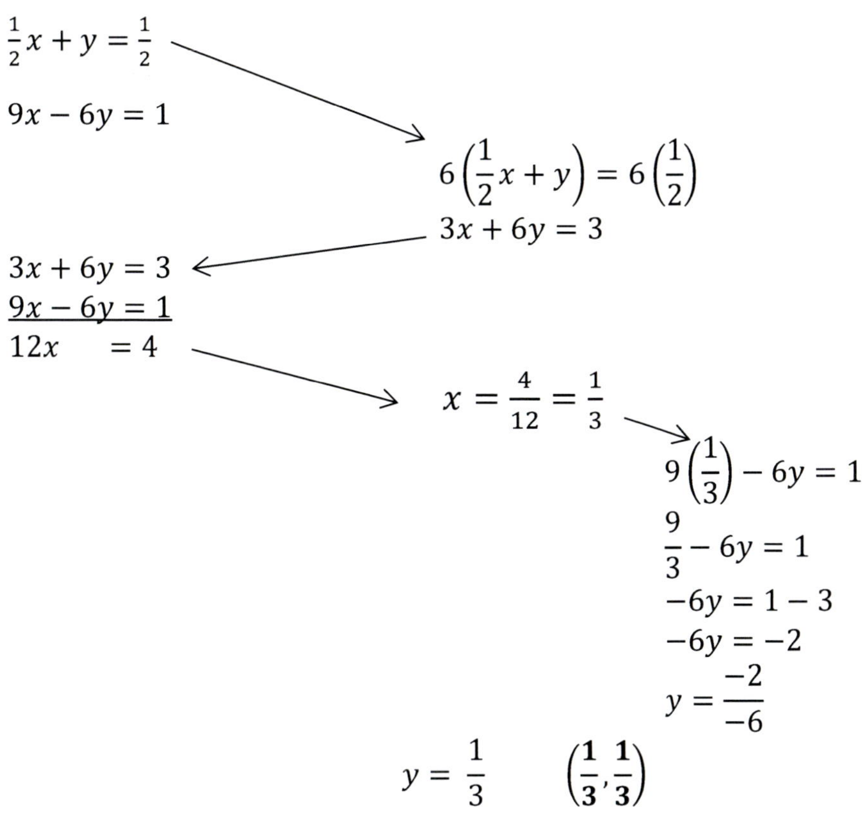

$\frac{1}{2}x + y = \frac{1}{2}$

$9x - 6y = 1$

$6\left(\frac{1}{2}x + y\right) = 6\left(\frac{1}{2}\right)$

$3x + 6y = 3$

$3x + 6y = 3$

$\underline{9x - 6y = 1}$

$12x \quad = 4$

$x = \frac{4}{12} = \frac{1}{3}$

$9\left(\frac{1}{3}\right) - 6y = 1$

$\frac{9}{3} - 6y = 1$

$-6y = 1 - 3$

$-6y = -2$

$y = \frac{-2}{-6}$

$y = \frac{1}{3} \qquad \left(\frac{1}{3}, \frac{1}{3}\right)$

By substitution:

$\frac{1}{2}x + y = \frac{1}{2}$

$9x - 6y = 1$

$y = \frac{1}{2} - \frac{1}{2}x$

$9x - 6\left(\frac{1}{2} - \frac{1}{2}x\right) = 1$

$9x - 3 + 3x = 1$

$12x = 1 + 3$

$12x = 4$

$\boldsymbol{x = \frac{4}{12} = \frac{1}{3}}$

$\boldsymbol{9\left(\frac{1}{3}\right) - 6y = 1}$

$\boldsymbol{3 - 6y = 1}$

$\boldsymbol{-6y = 1 - 3} \longrightarrow \boldsymbol{-6y = -2} \longrightarrow \boldsymbol{y = \frac{-2}{-6} = \frac{1}{3}} \qquad \boldsymbol{\left(\frac{1}{3}, \frac{1}{3}\right)}$

ANSWERS: Worksheet 11 page 4

4. By Addition:

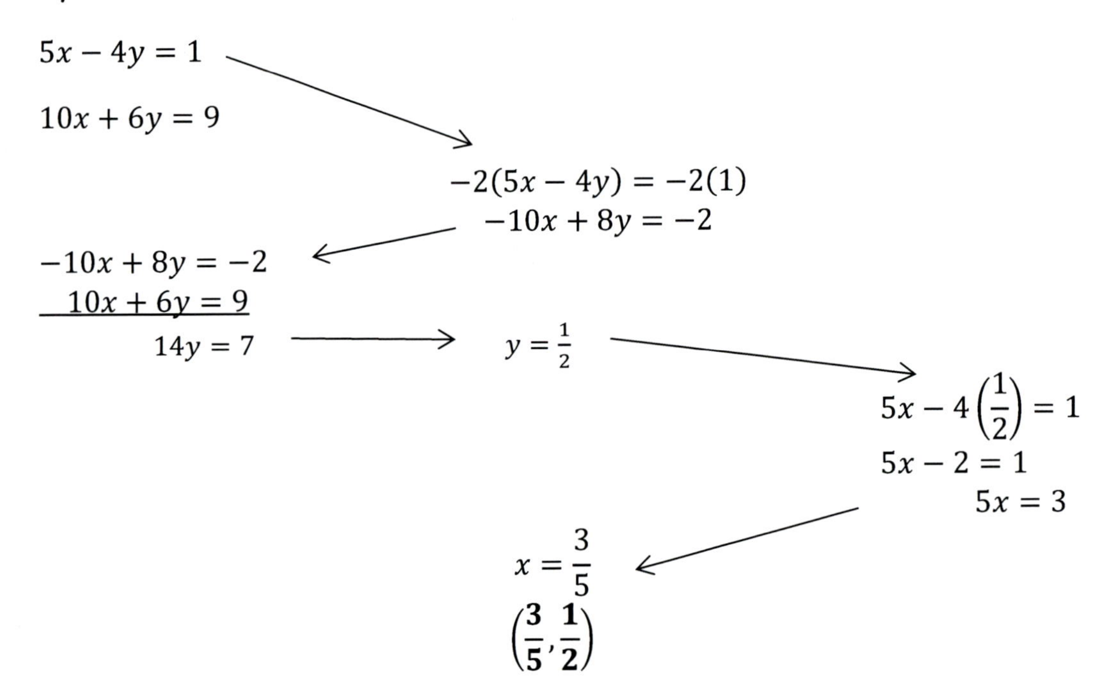

By Substitution:

$5x - 4y = 1$

$10x + 6y = 9$

$6y = 9 - 10x$

$y = \frac{9-10x}{6}$

$5x - 4\left(\frac{9-10x}{6}\right) = 1$

$5x - \left(\frac{4}{1} \cdot \frac{9-10x}{6}\right) = 1$

$5x - \left(\frac{36-40x}{6}\right) = 1$

$-\left(\frac{36-40x}{6}\right) = 1 - 5x$

Multiply both sides by -6 (cross cancel)

$\left(-\frac{6}{1}\right)\left(-\frac{36-40x}{6}\right) = (-6)(1 - 5x)$

$36 - 40x = -6 + 30x$

$36 + 6 = 40x + 30x$

$42 = 70x$

$\frac{42}{70} = x$

ANSWERS: Worksheet 11 page 5

$$x = \frac{6}{10} = \frac{3}{5}$$

$$5\left(\frac{3}{5}\right) - 4y = 1$$

$$\frac{15}{5} - 4y = 1$$

$$-4y = 1 - \frac{15}{5}$$

$$-4y = 1 - 3$$

$$-4y = -2$$

$$y = \frac{-2}{-4}$$

$$y = \frac{1}{2} \qquad \left(\frac{3}{5}, \frac{1}{2}\right)$$

ANSWERS: Worksheet 12

$$5x - 4y = 1$$
$$10x + 6y = 9$$

By Addition

$5x - 4y = 1$
$10x + 6y = 9$

$-2(5x - 4y) = -2(1)$

$-10x + 8y = -2$
$\underline{\;10x + 6y = 9\;}$
$+14y = 7$

$\boldsymbol{y = \frac{7}{14} \;\; = \frac{1}{2}}$

$10x + 6\left(\frac{1}{2}\right) = 9$

$10x + 3 = 9$

$10x = 9 - 3$

$10x = 6$

$x = \frac{6}{10} = \frac{3}{5}$

$\left(\frac{3}{5}, \frac{1}{2}\right)$

By Substitution

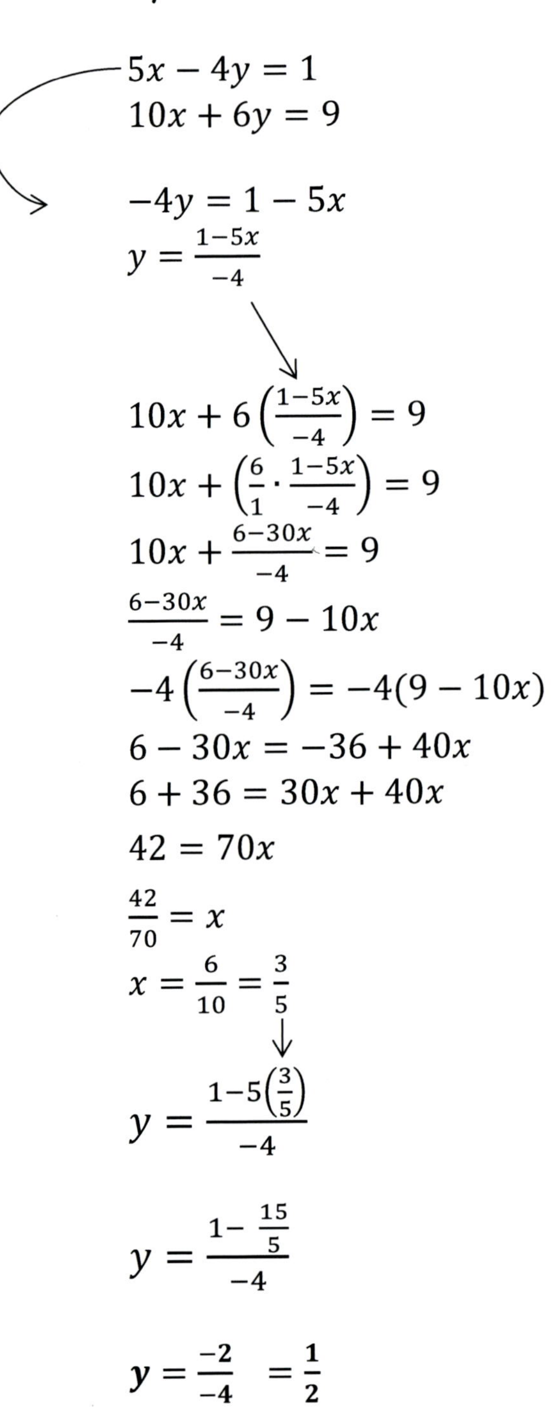

$5x - 4y = 1$
$10x + 6y = 9$

$-4y = 1 - 5x$

$y = \frac{1-5x}{-4}$

$10x + 6\left(\frac{1-5x}{-4}\right) = 9$

$10x + \left(\frac{6}{1} \cdot \frac{1-5x}{-4}\right) = 9$

$10x + \frac{6-30x}{-4} = 9$

$\frac{6-30x}{-4} = 9 - 10x$

$-4\left(\frac{6-30x}{-4}\right) = -4(9 - 10x)$

$6 - 30x = -36 + 40x$

$6 + 36 = 30x + 40x$

$42 = 70x$

$\frac{42}{70} = x$

$x = \frac{6}{10} = \frac{3}{5}$

$y = \frac{1-5\left(\frac{3}{5}\right)}{-4}$

$y = \frac{1-\frac{15}{5}}{-4}$

$\boldsymbol{y = \frac{-2}{-4} \;\; = \frac{1}{2}}$

ANSWERS: Worksheet 12 page 2

$$\left(\frac{3}{5}, \frac{1}{2}\right)$$

$5x - 4y = 1$

x	y
1	1
$\frac{3}{5}$	$\frac{1}{2}$
2	$\frac{9}{4}$

$5(1) - 4y = 1$
$-4y = 1 - 5$
$y = \frac{-4}{-4} = 1$

$5\left(\frac{3}{5}\right) - 4y = 1$
$\frac{15}{5} - 4y = 1$
$-4y = 1 - 3$
$y = \frac{-2}{-4} = \frac{1}{2}$

$5(2) - 4y = 1$
$10 - 4y = 1$
$-4y = -9$
$y = \frac{-9}{-4} = \frac{9}{4}$

$10x + 6y = 9$

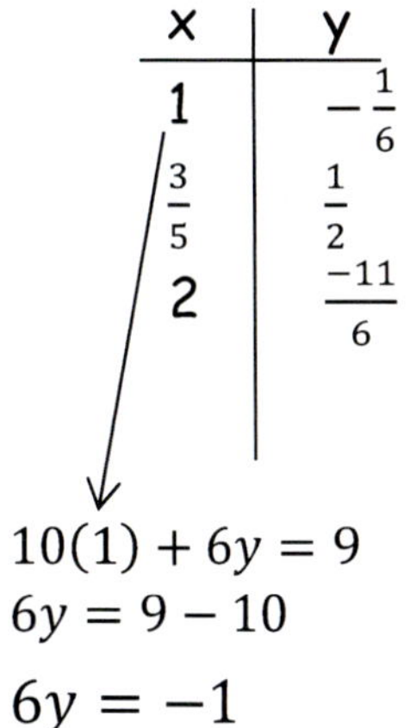

x	y
1	$-\frac{1}{6}$
$\frac{3}{5}$	$\frac{1}{2}$
2	$\frac{-11}{6}$

$10(1) + 6y = 9$
$6y = 9 - 10$
$6y = -1$
$y = -\frac{1}{6}$

$10\left(\frac{3}{5}\right) + 6y = 9$
$\frac{30}{5} + 6y = 9$
$6y = 9 - \frac{30}{5}$
$6y = 3$
$y = \frac{3}{6} = \frac{1}{2}$

$10(2) + 6y = 9$
$20 + 6y = 9$
$6y = -11$
$y = \frac{-11}{6}$

ANSWERS: Worksheet 12 page 3

$(1,1)\ \left(\frac{3}{5},\frac{1}{2}\right)\ \left(2,\frac{9}{4}\right)$

$\left(1,-\frac{1}{6}\right)\ \left(\frac{3}{5},\frac{1}{2}\right)\ \left(2,\frac{-11}{6}\right)$

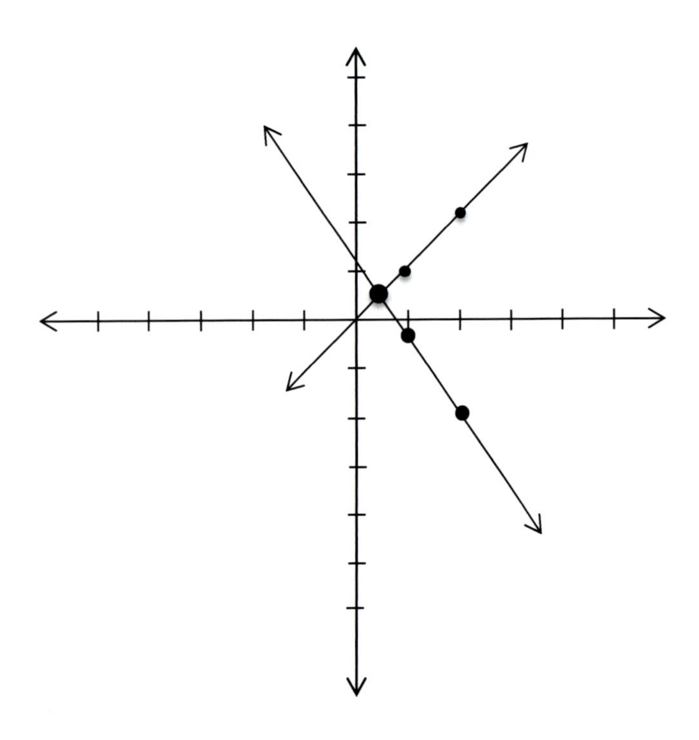

ANSWERS: Worksheet 13

1. The sum of two numbers is 43 and their difference is 9. Find the numbers.

$$\begin{array}{c} x + y = 43 \\ \underline{x - y = 9} \\ 2x = 52 \end{array}$$

$$\boldsymbol{x = 26}$$
$$\boldsymbol{y = 17}$$

2. Alvin is going to install some decorative tiles around a swimming pool. He will need to arrange 250 tiles using an assortment of plain white tiles and some expensive hand painted ones. The plain white tiles cost 50 cents each and the hand painted ones cost $3 each. He will charge $250 for his labor and he is charging $800 for the whole job. How many of each tile should he use to stay within the budget?

$$.5x + 3y + 250 = 800$$

$$x + y = 250$$

$$y = 250 - x$$

$$.5x + 3(250 - x) + 250 = 800$$

$$.5x + 750 - 3x + 250 = 800$$

$$-2.5x = 800 - 1000$$

$$-2.5x = -200$$

$$x = \frac{-200}{-2.5}$$

$$\boldsymbol{x = 80 \;\; and \; y = 250 - x \, , so \; y = 170}$$

ANSWERS: Worksheet 13 page 2

3. A plane is flying against the wind from Town A to Town B. The two towns are 500 miles apart. It took the pilot 4 hours get to Town B and 2 hours to get back to Town A. What were the speeds of the wind and the plane?

First Trip

$$\frac{500}{x-y} = 4$$

$$\frac{500}{4} = x - y$$

$$125 = \mathrm{x} - \mathrm{y}$$

Second Trip

$$\frac{500}{x+y} = 2$$

$$\frac{500}{2} = x + y$$

$$250 = \mathrm{x} + \mathrm{y}$$

$$125 = x - y$$

$$\underline{250 = \mathrm{x} + \mathrm{y}}$$

$$375 = 2x$$

$$\mathbf{187.50 = \boldsymbol{x}}$$

$$\mathbf{62.5 = \boldsymbol{y}}$$

The speed of the plane was 187.50 miles per hour. The wind was 62.5 miles per hour.

ANSWERS: Worksheet 14

1. The area of a garden is 48 square meters. If the length is 2 meters longer than the width, how much fencing will be required to enclose it?

$$Perimeter = 2(height) + 2(base)$$
$$Perimeter = 2(x) + 2(x + 2)$$

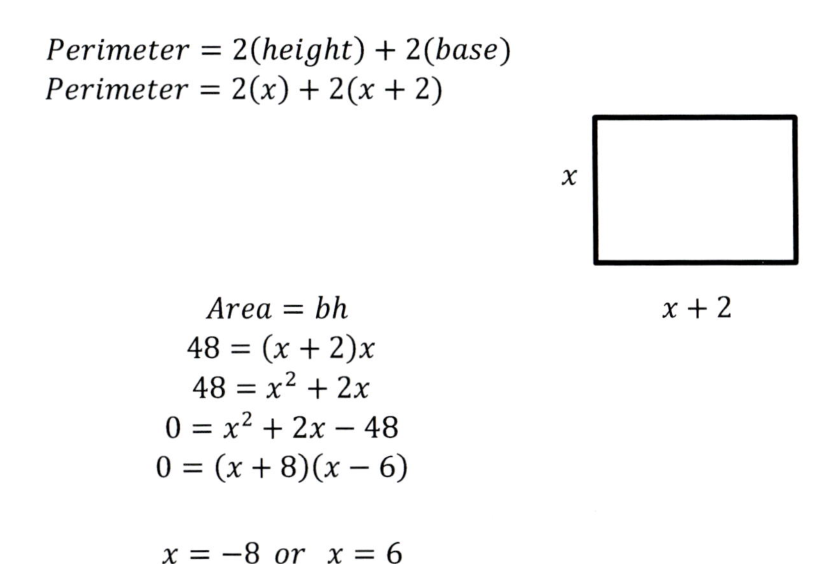

$$Area = bh$$
$$48 = (x + 2)x$$
$$48 = x^2 + 2x$$
$$0 = x^2 + 2x - 48$$
$$0 = (x + 8)(x - 6)$$

$$x = -8 \; or \; \; x = 6$$

Do you think the width is 6 or -8? I think it is 6, too. But the question is, "How much fencing will be required to enclose it?" **The answer is 28 meters.**

2. I need to order a box that is 4 inches deep with a volume of 320 in^3. The length needs to be 2 inches longer than the width. The factory would like to cut 4 inch squares out of the corners of a flat piece of cardboard and then flip up the sides to create the box. What should be the size of the flat piece of cardboard?

$$320 \, in^3 = 4 \, in \cdot x(x + 2)$$

$$320 = 4x(x + 2)$$
$$320 = 4x^2 + 8x$$

$$0 = 4x^2 + 8x - 320$$

ANSWERS: Worksheet 14 page 2

Divide each term by the coefficient of "a"

$$0 = \frac{4x^2}{4} + \frac{8x}{4} - \frac{320}{4}$$

$$0 = x^2 + 2x - 80$$

$$0 = (x + 10)(x - 8)$$

$$x = -10 \quad or \quad x = 8$$

I designed my equation to tell me the width of the box. The question asks for the dimensions of the flat cardboard. I will replace the x in our drawing with 8 inches and then add the two 4 inch cutouts. **That will make the width 16 inches and the length 18 inches.**

3. You need to fence off a rectangular shape next to a river. The area of the enclosure needs to be 96 square meters and the length should be 4 meters longer than the width. What should be the dimensions of the 3 fence sections?

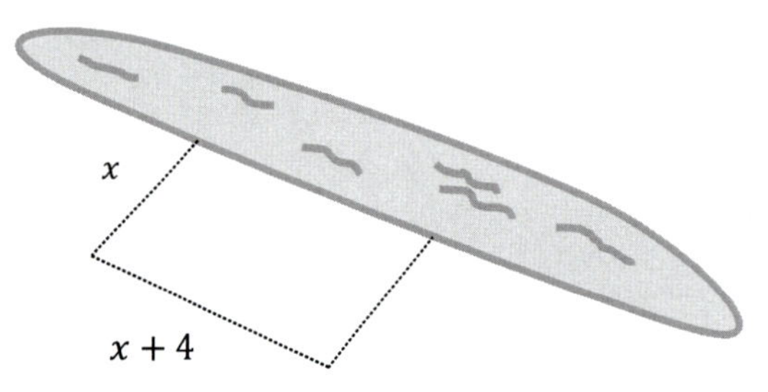

$$96 = x(x + 4)$$
$$96 = x^2 + 4x$$
$$0 = x^2 + 4x - 96$$

$$0 = (x + 12)(x - 8)$$
$$x = -12 \quad or \quad x = 8$$

Fence section #1 should be 8 meters.
Fence section #2 should be 12 meters.
Fence section #3 should be 8 meters.

ANSWERS: Worksheet 15

1. A man is going to row a boat up a stream at a steady pace for 4 miles. The rate of the current is 2 mph. He needs to row up the stream and then back down in 2 hours and 40 minutes. How fast should he row?

$$\frac{4}{x-2}+\frac{4}{x+2}=\frac{8}{3}$$

Common Denominator
$(x-2)(x+2)(3)$

$$\frac{4}{\cancel{x-2}}\cdot\frac{\cancel{(x-2)}(x+2)(3)}{1}+\frac{4}{\cancel{x+2}}\cdot\frac{(x-2)\cancel{(x+2)}(3)}{1}=\frac{8}{\cancel{3}}\cdot\frac{(x-2)(x+2)\cancel{(3)}}{1}$$

$$[4(x+2)\cdot 3]+[4(x-2)\cdot 3]=8(x-2)(x+2)$$

$$12(x+2)+12(x-2)=8(x^2-4)$$

$$12x+24+12x-24=8x^2-32$$

$$24x=8x^2-32$$

$$0=8x^2-24x-32$$

$$0=\frac{8x^2}{8}-\frac{24x}{8}-\frac{32}{8}$$

$$x^2-3x-4=0$$

$$(x+1)(x-4)=0$$

$$x=-1 \qquad or \qquad x=4$$

The man should row 4 miles per hour.

ANSWERS: Worksheet 15 page 2

2. In the last problem, you ended up with an equation that was in the standard for of a quadratic equation. Write that equation here:

$$x^2 - 3x - 4 = 0$$

Use that equation to fill in the table below and then plot those points onto the graph on the next page.

$f(-2)$	$f(-1)$	$f(0)$	$f(1)$	$f(2)$	$f(3)$	$f(4)$	$f(5)$
6	0	-4	-6	-6	-4	0	6

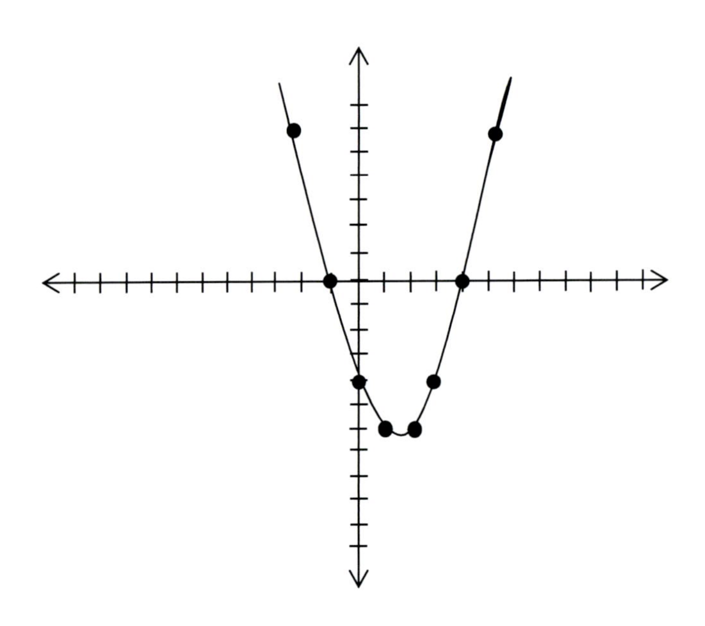

ANSWERS: Final Test

1. A race car driver is trying to beat his opponent's record. His opponent drove 200 miles in 1 hour and 10 minutes. If the race car driver travels at an average speed of 175 miles per hour, will he beat his opponent's record?

$$\frac{200\ miles}{175\ mph} = x$$
$$200 \div 175 = 1.14$$

How long is that? Did he break the record?

1.14 of 60 minutes equals

$$1.14 \times 60 = 68.4\ minutes$$

He will finish the race in $\boldsymbol{68.4\ minutes}$. He will break the record!

2. Esther jogged at 4 mile per hour for 2 hours and 45 minutes. How far did she get?

$$\frac{x}{4} = \frac{11}{4}$$

$$4x = 44$$

$$\boldsymbol{x = 11\ miles}$$

ANSWERS: Final Test page 2

3. Teresa and Jessica are sewing some aprons. Teresa has been sewing for many years and she can complete 3 aprons in just 1 hour. Jessica is new at this job and can only sew 1 apron in 2 hours. If the two of them work together for 8 hours, how many aprons will they make?

$$\frac{x}{\frac{3}{1}+\frac{1}{2}}=8$$

$$\frac{x}{\frac{6}{2}+\frac{1}{2}}=8$$

$$\frac{x}{\frac{7}{2}}=8$$

$$x=\frac{8}{1}\cdot\frac{7}{2}$$

$$x=\frac{56}{2} \qquad \boldsymbol{x=28\ aprons}$$

4. Steve, Mike, and Eric worked for 6 hours and 40 minutes cutting grass. After the first hour, Steve had mowed $\frac{3}{4}$ of the first lawn. Mike had mowed $\frac{1}{2}$ of the second lawn and Eric had mowed $\frac{1}{4}$ of the third lawn. If they keep up this same pace for 6 hours and 40 minutes, how many lawns will they mow?

$$\frac{x}{\frac{3}{4}+\frac{1}{2}+\frac{1}{4}}=\frac{20}{3}$$

$$\frac{x}{\frac{3}{4}+\frac{2}{4}+\frac{1}{4}}=\frac{20}{3}$$

$$\frac{x}{\frac{6}{4}}=\frac{20}{3}$$

$$x=\frac{20}{3}\cdot\frac{6}{4} \qquad x=\frac{120}{12} \qquad \boldsymbol{x=10\ lawns}$$

ANSWERS: Final Test page 3

5. The two triangles below are congruent. Solve for x.

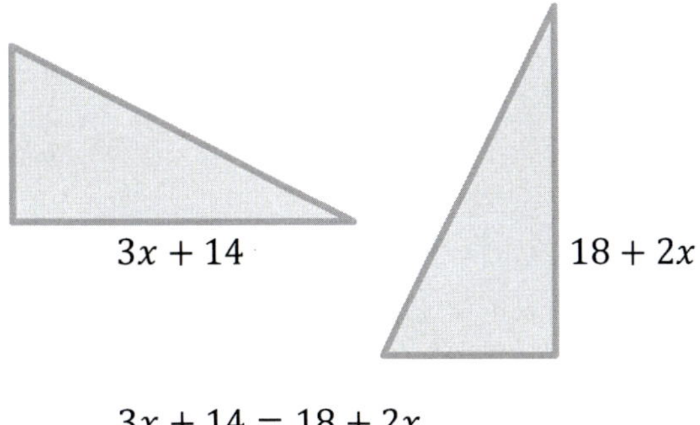

$3x + 14 = 18 + 2x$
$3x - 2x = 18 - 14$
$\boldsymbol{x = 4}$

6. What is the area of the shaded triangle below?

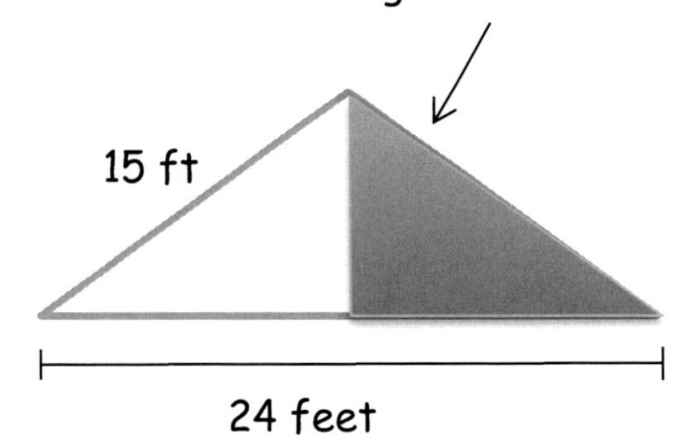

To find the area, we will need to know the height. The blue triangle and the white triangle are congruent, so the blue triangle has the same dimensions. I will use the Pythagorean Theorem to solve for the height of the blue triangle.

$$x^2 + 12^2 = 15^2$$
$$x^2 + 144 = 225$$
$$x^2 = 225 - 144$$
$$x^2 = 81 \quad x = 9$$

The height of the triangle is 9 feet. I will solve for the area of just the blue triangle.

$$A = \frac{1}{2}(12)(9)$$

$$\boldsymbol{Area = 54\ ft^2}$$

ANSWERS: Final Test page 4

7. A builder needs to re paint the green triangular portion of this house a different color. He is told that the roof has a pitch of 6:12 and he knows that the blue board is 32 feet long. Each can of paint will cover 140 square feet. How many cans of paint will he need?

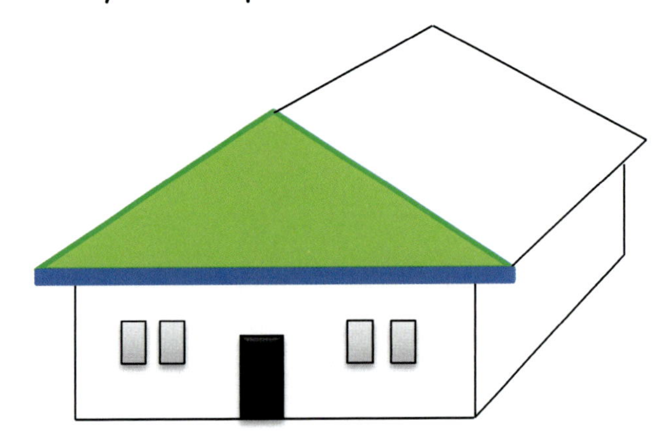

You need to find the area of the green triangle. The pitch of the roof is 6:12. We will use that similar triangle to find the missing dimensions.

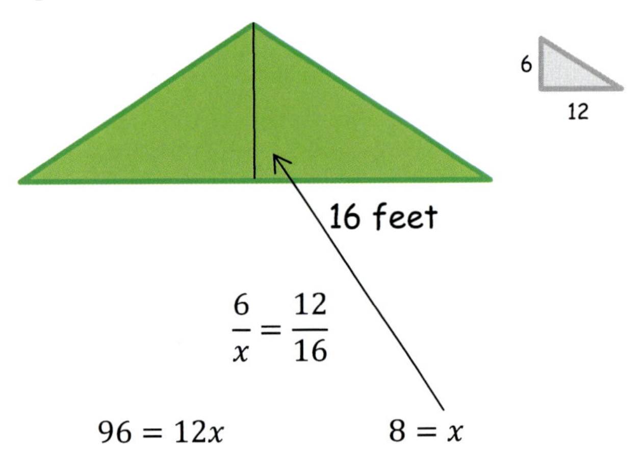

$$\frac{6}{x} = \frac{12}{16}$$

$$96 = 12x \qquad 8 = x$$

$$Area\ to\ be\ painted = \frac{1}{2}(32)(8) \qquad Area = 128\ ft^2$$

The builder needs only 1 can of paint.

ANSWERS: Final Test page 5

8. What is the probability of rolling an even number on a regular 6-sided die?

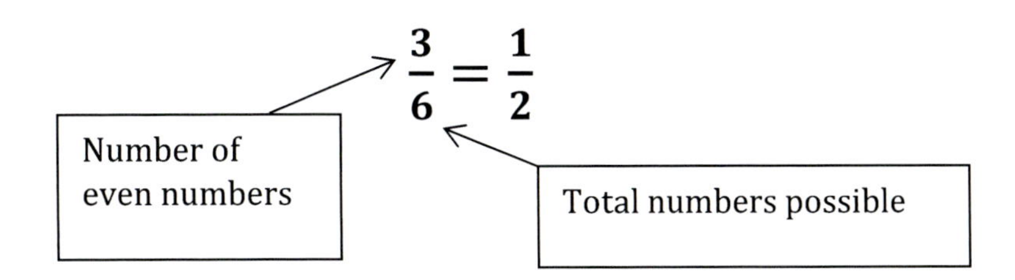

9. In a carton of a dozen eggs, 3 of the eggs are broken. You randomly select 2 eggs to cook. What is the probability of selecting 2 broken eggs?

$$\frac{3}{12} \cdot \frac{2}{11} = \frac{6}{132} \qquad = \mathbf{\frac{1}{22}}$$

10. A bowl contains 5 girl names and 7 boy names. What is the probability of randomly selecting two boy names or two girl names?

Probability of selecting 2 girl names →

$$\frac{5}{12} \cdot \frac{4}{11} = \frac{20}{132}$$

$$\frac{7}{12} \cdot \frac{6}{11} = \frac{42}{132}$$

← Probability of selecting 2 boy names

$$\frac{20}{132} + \frac{42}{132} = \frac{62}{132} \qquad = \mathbf{\frac{31}{66}}$$

Probability of selecting 2 boy **OR** 2 girl names

ANSWERS: Final Test page 6

Solve the inequalities below.

11. $3(x + 2) < 5(x - 4)$

$$3x + 6 < 5x - 20$$
$$6 + 20 < 5x - 3x$$
$$26 < 2x$$
$$\mathbf{13 < x}$$

12. $\frac{x}{5} - \frac{x}{2} < 3$

$$\frac{2x}{10} - \frac{5x}{10} < 3$$

$$\frac{-3x}{10} < 3$$

$$-3x < 3 \cdot 10$$

$$-3x < 30$$
$$\frac{-3x}{-3} > \frac{30}{-3}$$

$$\boldsymbol{x > -10}$$

13. $-16x < -32$

$$x > \frac{-32}{-16}$$

$$\boldsymbol{x > 2}$$

14. $-\frac{2}{5}y < 20$

$$y > \frac{20}{-\frac{2}{5}}$$

$$y > \frac{20}{1} \times -\frac{5}{2}$$

$$\boldsymbol{y > -50}$$

ANSWERS: Final Test page 7

15. $5y - 7 > 4 - y$

$$5y + y > 4 + 7$$
$$6y > 11$$
$$y > \frac{11}{6}$$

16. You are told to create 50 cm^3 of a solution that is 8% bleach and water. How much bleach should you add to what amount of water?

$$1x + 0y = .08(50)$$
$$x + y = 50$$
$$x = 50 - y$$

$$1(50 - y) + 0y = .08(50)$$
$$50 - 1y + 0 = 4$$
$$-1y = -46$$
$$y = 46$$

You should mix 4 cm^3 of bleach with 46 cm^3 of water.

17. You are instructed to fill a fish tank with 400 gallons of saltwater that is 25% salt. You are given two containers of different saltwater. One is 40% salt the other one is 20% salt. How much of each solution should you pour into the tank?

$$.4x + .2y = 400(.25)$$

$$x + y = 400$$
$$y = 400 - x$$

$$.4x + .2(400 - x) = 400(.25)$$
$$.4x + .80 - .2x = 100$$
$$.2x = 100 - 80$$
$$.2x = 20$$
$$x = \frac{20}{.2} \qquad x = 100$$

Add 100 gallons of the 40% saltwater and 300 gallons of 20%

ANSWERS: Final Test page 8

18. A 100 cm^3 solution of acid and water is 26% acid. How many cubic centimeters of pure water should be added to make a solution that is 13% acid?

$$100(.26) + 0(x) = .13(y)$$
$$100 + x = y$$
$$y = 100 + x$$

$$100(.26) + 0(x) = .13(100 + x)$$
$$26 + 0 = 13 + .13x$$
$$26 - 13 = .13x$$
$$13 = .13x$$
$$x = \frac{13}{.13} \quad x = 100cm^3$$
$$\boldsymbol{x = 100\ cm^3\ of\ pure\ water}$$

19. A candy store wants to sell 4-pound bags of candied popcorn and peanuts at a price of $9.60 cents per pound. The candied popcorn costs 50 cents per ounce and the peanuts cost 65 cents per ounce. How many ounces of each ingredient should be mixed together to create such a mixture?
(Remember to keep your units the same).

A pound equals 16 ounces, so the candy will be sold in 64-ounce bags at 60 cents per ounce.

$$x = ounces\ of\ popcorn$$
$$y = ounces\ of\ peanuts$$
$$x + y = 64\ ounces$$

$$.5x + .65(64 - x) = 64(.6)$$
$$.5x + 41.6 - .65x = 38.4$$
$$-.15x = 38.4 - 41.6$$
$$-.15x = -3.2$$

$$x = 21.33$$
$$21.33 + y = 64$$

$$y = 42.67$$

The store owner should add 21.33 ounces of popcorn and 42.67 ounces of peanuts.

ANSWERS: Final Test page 9

20. A street vendor is selling balloons. He decides to create some bouquets of 15 balloons and sell them for $20 each. The regular latex balloons full of helium cost $1 each. The shiny Mylar balloons full of helium cost $1.50 each and the string, ribbons, and fancy weight cost $2 for each bouquet. How many of each balloon should he use?

$$x = latex\ balloons \qquad y = Mylar\ balloons$$

$$x + y = 15$$

$$1x + 1.5(15 - x) + 2 = 20$$

$$x + 22.5 - 1.5x + 2 = 20$$

$$-.5x + 24.5 = 20$$

$$-.5x = -4.5$$

$$\boldsymbol{x = 9} \qquad \boldsymbol{y = 6}$$

The street vendor should use 9 latex balloons and 6 Mylar balloons.

21. Solve for the simultaneous solution by using the addition method.

I multiplied both sides by -3

$$x + 2y = -1$$
$$\underline{3x + 5y = -4}$$

$$-3(x + 2y) = -1(-3)$$
$$-3x - 6y = 3$$
$$\underline{3x + 5y = -4}$$
$$-y = -1$$
$$\boldsymbol{y = 1} \qquad \boldsymbol{x = -3}$$

ANSWERS: Final Test page 10

22. Solve for the simultaneous solution by using the substitution method.

$$x + 4y = 13$$
$$4x - y = 18$$

$$x + 4y = 13$$
$$x = 13 - 4y$$

$$4(13 - 4y) - y = 18$$
$$52 - 16y - y = 18$$

$$-17y = 18 - 52$$

$$-17y = -34$$

$$\mathbf{y = 2} \qquad \mathbf{x = 5}$$

23. A plane flew from point A to point B against the wind for 4 hours. The plane returned back to point A in 3 hours. The distance between point A and point B is 1200 miles. What were the rate of the wind and the speed of the plane?

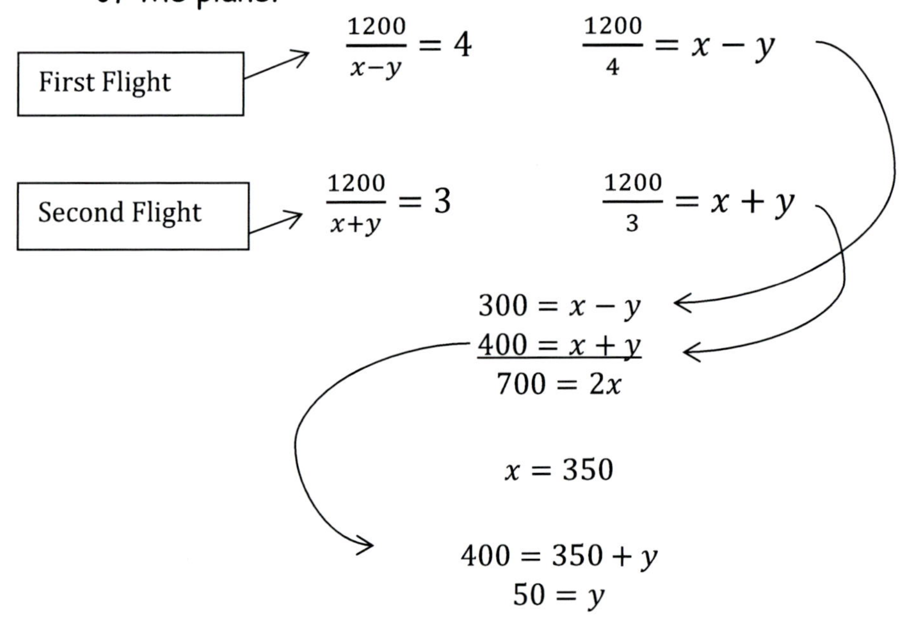

The rate of the plane was 350 mph and the rate of the wind was 50 mph.

ANSWERS: Final Test page 11

24. A train is traveling east at 80 mph. Another train is traveling west at 100 mph, just 90 miles away. How long will it take for the trains to meet and how far will each train have traveled?

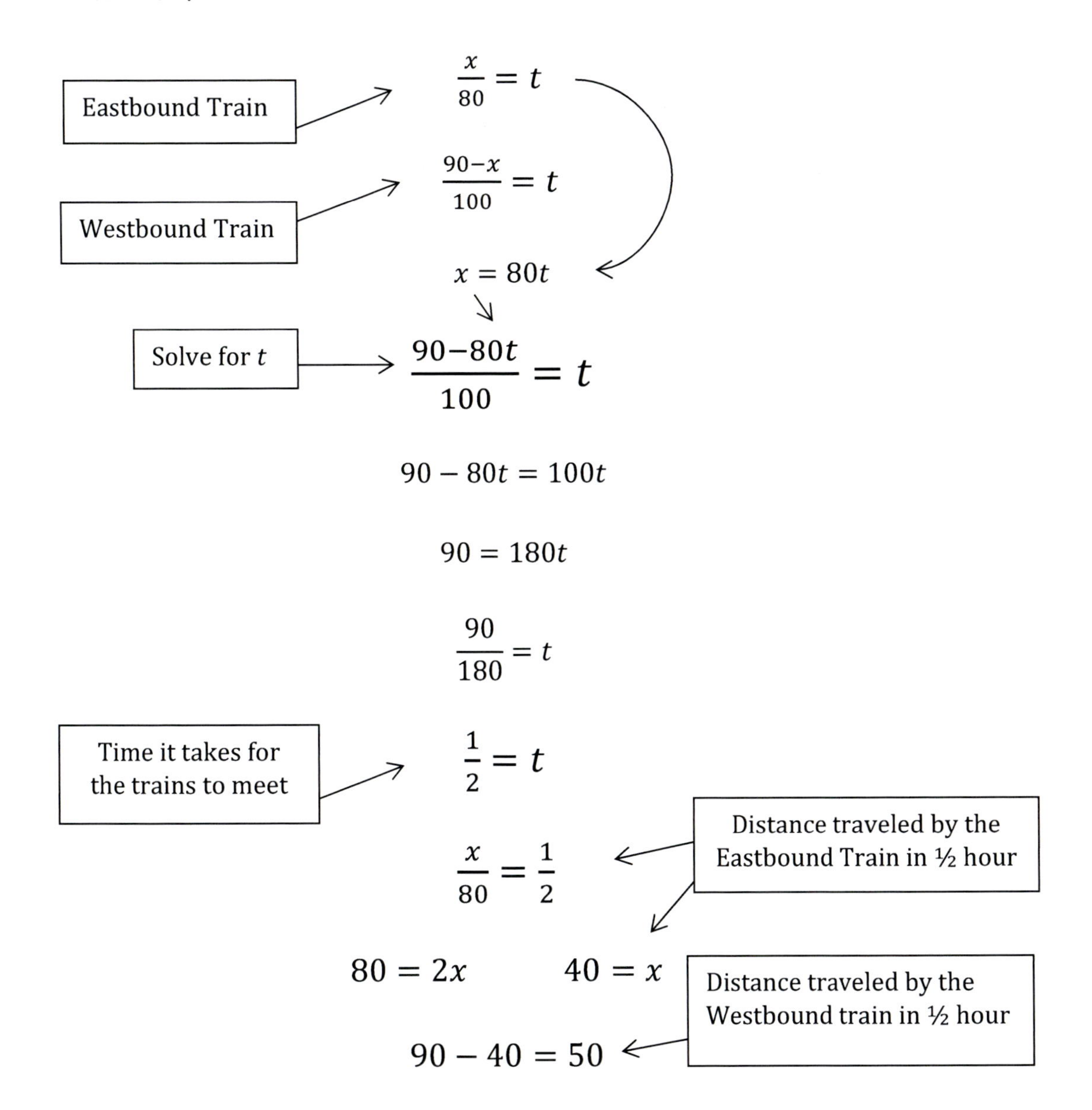

The trains met in 30 minutes. The Eastbound train had traveled 40 miles and the Westbound train had traveled 50 miles, when the two trains met.

ANSWERS: Final Test page 12

25. A woman is trying to swim 2 miles upstream against the current and then back downstream in 30 minutes. If the rate of the current is 3 miles per hour, how fast should she swim?

$$\frac{2}{x-3}+\frac{2}{x+3}=\frac{1}{2}$$

$$2(2)(x+3)+2(2)(x-3)=1(x-3)(x+3)$$

$$4x+12+4x-12=(x^2-9)$$

$$8x=x^2-9$$

$$x^2-8x-9=0$$

$$(x+1)(x-9)$$

$$x=-1 \quad or \quad x=9$$

The woman should swim at 9 mph. She better get some flippers!

ANSWERS: Final Test page 13

26.In the last problem, you created a quadratic. Write it below.

$\underline{x^2 - 8x - 9}$

Fill in the table below and then plot those points onto the graph.

$f(-2)$	$f(0)$	$f(2)$	$f(4)$	$f(5)$	$f(6)$	$f(8)$	$f(10)$
11	-9	-21	-25	-24	-21	-9	11

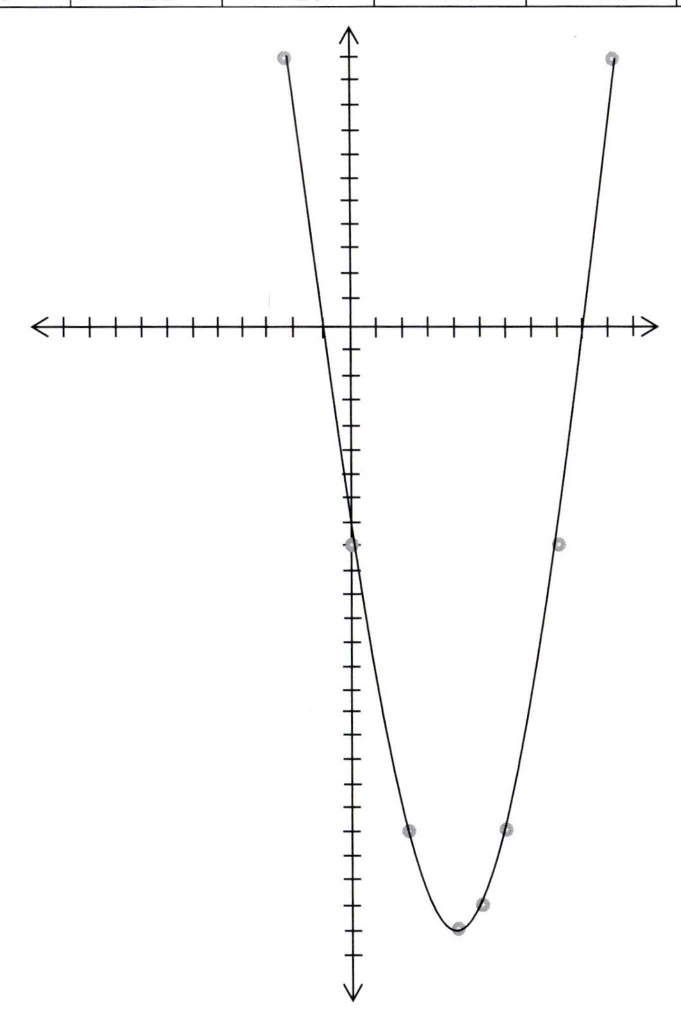

Made in the USA
Monee, IL
09 April 2022